SEAL WARS

BY

JANICE SCOTT HENKE

BREAKWATER

©1985 Janice Scott Henke

Canadian Cataloguing in Publication Data

Henke, Janice Scott
    Seal wars: an American viewpoint

Bibliography
ISBN 0-919519-61-X (bound) — ISBN 0-919519-63-6 (pbk)

1. Sealing — Newfoundland.   2. Sealing — Newfoundland — Public opinion.   I. Title.
SH362.H46 1985   639'.29'09718   C85-099217-6

*Cover photo Brian Roberts;
inserts Janice Henke*

Breakwater gratefully acknowledges the support of The Canada Council.

CAUTION

No part of this publication may be photocopied, or otherwise reproduced, without the Publisher's permission: Breakwater Books Ltd., 277 Duckworth Street, St. John's, Newfoundland, A1C 1G9.

Contents

Introduction		9
I	Save The Seals! The Protest Movement Wages a Multimedia War	11
II	Seal Time	18
III	The Way It Was	37
IV	To Kill A Seal: Development of Agreement on Humane Technique	47
V	Saving the Seals: History of Seal Management in Canada	63
VI	The Seal Saviours: Organizations of Protest	81
	The Animal Protection Institute	82
	Friends of Animals	86
	The Fund for Animals	88
	The Humane Society of the United States	91
	The Seal Rescue Fund	97
	Greenpeace	102
	The Sea Shepherd Conservation Society	124
	The International Fund for Animal Welfare	139
VII	Victims of the Seal Wars	175
Postscript		207
Bibliography		213

Acknowledgements

This book has been a research project which has spanned four years. During that time, all manner of people who have cared, either privately or professionally, about harp seals and their welfare, have generously contributed to my knowledge. My final impressions, however, are the result of my own synthesis of all this information. Needless to say, it will not please everyone. Yet, most will probably admit that there has been an attempt to present all sides of this issue. The degree of objectivity achieved is debatable. This is a subject on which no feeling, compassionate human can remain entirely neutral. My hope is that a greater amount of information offered here in this manner, shall stimulate constructive thought and begin a process of re-evaluation of our place in a world where predation is still a natural element.

Those who have contributed to my knowledge of general seal biology, physiology, breeding behaviour, feeding patterns and migration behaviour include: Dr. Keith Ronald and his colleagues at the University of Guelph, Ontario; Dr. Dave Sergeant of the Arctic Biological Station at Ste. Anne de Bellevue, Quebec; and Dr. Dave Lavigne of the University of Guelph. These scientists' published works on the harp seal are the basis for all the detailed information in "Seal Time," and if the present interpretation of seal life is flawed, the fault is mine, not theirs. I thank Dr. Sergeant for his comments and critique of that chapter in its formative stages.

The work of Dr. W. Don Bowen of the Research and Resource Services section of the Department of Fisheries and Oceans in St. John's, Newfoundland, and that of Dr. Sergeant, added to my understanding of the problems and progress in seal tagging, recapture statistics, and the growing science of prediction of pup production and seal population estimates. Statements pertaining to the status of the harp seal in the western Atlantic are based in this and other research and on the criticisms of it which have been made by the side of protest. Suffice it to say that I am glad no one knows exactly how many seals are out there: if we did, they would be in trouble. Population and production estimates are being increasingly substantiated by field check techniques.

My knowledge of the seals would be woefully inadequate if I had never had the opportunity to see, touch and hear them for myself. I suspect that if it had not been for Tom Hughes, Paul Montreuil and Brian Roberts I might not have obtained permission to go. Thank you, Gentlemen, for believing that I wished to make an honest effort in reporting on the realities of the hunt.

My first visit to the ice was greatly enhanced by the courtesy and consideration of the law enforcement and biological research teams of the Department of Fisheries and Oceans stationed in the Gulf on the Magdalen Islands. Biologist Ron Greendale and fisheries officer Wyb Hoek introduced me to harp and hooded seals during the spring of 1982 when I also observed the last commercial hunt there, thanks to the cooperation of chief officer Roger Simon. My memories and photographic opportunities were a once-in-a-lifetime experience. I thank them with all my heart.

William McCloskey has, through his thorough and candid accounts of sealing and seal hunt protesters, earned the respect of many people. Although we have never met, nor even had a telephone conversation, we have corresponded briefly on the subject of this book. I thank him sincerely for his advice and admire his own courageous perspective in this matter. It may be said that Bill was among the first to publically dispute the unfounded claims of protest. His list of publications is an indication that his words have a deserved credibility.

Although I had been told that I would never receive straight answers from Canadian government officials, I have never been refused a prompt answer to any question which was asked, nor have I ever felt that any Department of Fisheries and Oceans personnel displayed a reluctance to discuss anything in relation to the seal hunt controversy. In fact, much of my access to seal hunt protest literature came originally from the DFO, which also generously allowed me to go through letters of protest which had been collected for two decades. Both protest films and documentaries were offered by the Department for my review, as were all manner of photographs of the hunt and newspaper accounts of protest.

All sides of the PEI massacre were made known to me voluntarily by the government of Canada. A frank discussion of why it had happened convinced me that human error, not cruel or routine mismanagement, had been a contributing factor. Much was learned in those sad days in the spring of 1981.

The history of international involvement in harp and hooded seal management has been enhanced by the generosity of the Department of Fisheries and Oceans, whose personnel have continuously supplied me with major research papers such as that of Drs. Beddington and Williams, the ICES Working Group Report of 1982, the report of the Nature Conservancy Council of 1982, the Canadian critique of it which was presented to the CITES convention in Botswana in 1983, and The North Atlantic Fisheries Organization "Assessment of Seal Stocks" in the Council reports of 1983.

It has to be said that no piece of research which might have been withheld, ever was. All manner of published information, both supportive of Canada's seal management, and highly critical of it, was offered to me as relevant. For this I shall always thank Brian Roberts, Paul Montreuil, Dan Goodman, Peter Meerburg and Jim Winter. They have my highest respect and admiration.

Each of the organizations of protest which I have reported upon has generously sent me its policy statements and informational literature on the harp seal issue. I thank them for this cooperation and especially wish to thank certain ones for their extra measure of effort in seeing to it that I understood their continuing efforts at ending the hunt. Cleveland Amory, of the Fund For Animals, took the trouble to call me personally with his opinions. I shall never forget that particular interview.

Alice Herrington of the Friends of Animals sent her organization's policy statement. I am grateful for having been continued on the Friends mailing list.

Brian Davies and Steve Best of the International Fund For Animal Welfare have sent personal notes, Davies' book *Seal Song*, spoken with me on the telephone, and kept me informed through inclusion on their mailing list, and with position papers on the seal hunt at the time of the European pelt ban vote and the fish boycott. Without their generous cooperation, my research would have been flat, indeed.

Greenpeace USA, the Seal Rescue Fund, the Humane Society of the United States, and the Animal Protection Institute of America have generously sent me their packets of seal hunt protest literature, without which very little could have been said about their activities in this regard.

I wish to thank Investigator Jerry Conway for the loan of the children's letters to Paul Watson, which were found on the *Sea Shepherd II*. They helped greatly to illustrate the effect of hero worship which has been a part of the process of protest.

Mr. Tom Hughes of the Ontario Humane Society and the Committee on Seals and Sealing kindly spoke with me about the humane killing of seals and his work with enthanasia concerns in general. This gentleman should always be remembered in Ontario for his impact on the laws governing the humane slaughter of domestic stock and his concern that the seal hunt should always be properly conducted. I thank him for his time and courtesy, and respect him for his candor and intellectual honesty.

I thank Jim Gourlay of the Halifax *Chronicle Herald* for a thorough introduction to the scope of the seal hunt issue, and for his time in briefing me on conditions in the Gulf of St. Lawrence prior to my visit there.

Thanks also to Judy Swan of the Department of Fisheries and Oceans, for her enthusiasm, research tips, and encouragement in the culmination of this project.

My friends and acquaintances in Newfoundland who have shown me great hospitality and helped me understand the importance of both the offshore and the landsmen's hunt to their total cultural and economic life shall never be forgotten. Thank you Roy and Grethe Pilgrim, Francis Patey, Ralph Rose, Captain and Mrs. Morrissey Johnson, Mark Small, Jack, Florence, Gary and Mr. Troake, Cal and Kay Waterman, Ches and Bernice Coish, Stan and Jean Winter, Jim and Sharon Winter, and Kirk and Ulrike Smith. Thank you to the Canadian Sealers Association for adding to my research and smoothing my introductions to all those who were willing to tell still another writer from away about their life with and without the seals.

Thank you to Clyde Rose, publisher, and Dick Buehler, editor, for having the patience and courage to print my words, and better organize my ideas. I feel we have accomplished something important together in telling this story in this manner for the first time.

Thank you Bob, my loving husband, for your continuous support and faith that I could and should finish this book. Because of you, I have realized great personal growth and development through exploring new ideas and experiences. What more could one ask of a life mate?

Introduction

This book is written for those who have been affected by the "Save the Seals" protest, to aid them in understanding how it came about and what "animal welfare" really is. The growth and impact of those organizations which claim to have saved the harp seals is examined, as is the conservation programme of the government of Canada.

Those people who have been victimized by the Seal Wars of the twentieth century are viewed in their widely separated camps; they live in modern Europe, in America, and in the fishing villages of Canada and Greenland. So far there has been no meaningful dialogue, and no truce, between them and the various protest organizations.

The campaigns in this war have been staged by public relations experts, enhanced by television and newspaper coverage of well-orchestrated demonstrations, and financed by millions of dollars donated in good faith by good people.

The following account is an accurate one. It describes harp seals as they are: beautiful, wary, abundant and specialized fish eaters which migrate yearly from the northern Arctic to more southerly waters to bear their young.

The story is also about seal hunters and their families. Inuit of Canada and Greenland, who have always depended on seals for food and income, find their destinies manipulated from thousands of miles away by those who care little for their miserable poverty and new economic confusion. Newfoundlanders and other Maritimers who have hunted seals for four hundred years cannot understand why outsiders hate them so, and why they refuse to believe that the seals are as important to them as the fish, which are the staff of life.

Those who have wanted to save seals from hunters deserve special attention. Their beliefs have been manipulated, unfairly coloured, and used in schemes of economic blackmail unprecedented in the history of modern protest. The spirit of Earth Day has carried with it an unwavering trust in the "New Ecologists," those cultural heroes who have done their best to expose the dangers of air pollution, over-exploitation of resources, nuclear arms proliferation, acid rain, and toxic waste to a world which has finally slowed down enough to think about these things. The heroes' messages, enhanced by each other, have been believed and heeded. Who can now doubt the motives or the truth of a statement made by Greenpeace? Why would they and the International Fund For Animal Welfare be screaming to the world about the dangers of hunting seals if their warnings were not well-founded? Weren't they right about the whales, many of which are now on endangered lists? Dare we ignore their warnings? Or has our trust been unjustified?

The reasons for the growing credibility of the protest movement against the hunting of harp seals, are a part of the Seal Wars story. This book is dedicated to all those who have feared for and loved the seals in the hope that increased knowledge will make their love more meaningful, and their response more constructive.

I

Save The Seals!
The Protest Movement Wages
A Multimedia War

Scenario: North America
Time: March 1980

"My God! Look at this! They're hitting those babies with clubs and then skinning them alive! There it is again! It's unbelievable — so cruel.... Those murdering creeps ought to be shot!"

The man who was being spoken to nodded weakly and put down his steak-filled dinner fork. The television was a wonderful invention, he mused. Here he was, sitting in comfort trying to eat a decent meal while a thousand miles away men were out in the Atlantic beating seals to death while he watched. They always had these stories on during dinner, too. The seal corpse on the screen glowed as red as anything he had ever seen, and it twitched in a convulsive fashion. As steam rose from the body, the camera panned back to sweep the entire area. Jagged sea ice was spotted with red blots and red drag marks.

More little white seals were being clubbed to death by men dressed in dark snowmobile suits. The clubbing was rhythmical, three blows to each skull, and when the man finished hitting the seal at his feet he straightened and took his knife from his belt. He bent down, rolled the body over onto its back and thrust the knife deep into the chest, between the front flippers. Blood gushed onto the ice. Within seconds, the knife slid down the underside of the body, from chin to tail. With quick, deft strokes, the man separated white blubber from the red meat below, cutting neat arm holes where the front flippers were. He separated the skin with its heavy load of fat from the sides, the back, and the rear flippers and tail, and flung the pelt onto the ice, fat side up. It steamed, the carcass steamed, and the man straightened up and looked around for another pup. He glared at the camera, and turned his back, striding off camera.

"Enough already," muttered the man who had stopped eating. But the scene continued, the narrator droning on: "...and this year, some 150,000 baby harp seals will be slaughtered in the annual spring harvest of these animals for their fur. As usual, Greenpeace is here. A number of these volunteer protesters have been arrested on the ice for trying to block the sealers' blows with their own bodies as the season opened on a beautiful March day. The Canadian government has sent its mounties and fisheries officers out to the ice to prevent the annual protest and the interference with the sealing operation, but as usual, Greenpeace members have managed to get through and make their opinions on this hunt known to the world. They claim many seals are actually skinned alive and that the species is in danger of extinction if this hunt continues."

Again, relentlessly, the camera zoomed in to a seal which had just been struck. Blood poured from its nose and mouth as its head dropped to the ice. Its eyes remained open as the club descended again and again, crushing the skull.

"Jesus Living Christ, turn it off!" But the man and his wife, neither of whom had ever seen anything butchered, much less performed the deed personally, continued to stare numbly at their television set until the commercial started, with its music and its laxative message.

"Sam Peckinpah's got nothing on those guys," said the man to his wife. They agreed that something should be done to stop that atrocity once and for all.

In homes all across America, Great Britain, Europe and Australia, similar scenes were repeated. Millions of viewers recoiled from the sight of the actual slaughter of seal pups. Reaction varied from shock to outrage to mild interest.

"Same old thing every year," said one man, who took this excuse to head for the bathroom when the feature first came on. He felt it made dull news, especially since he had a black and white set. He had seen it times enough before.

Televised news broadcasts were not the only avenues by which the "New Ecologists" reached the public with "information" about the harp seal hunt. Entertainment feature programmes such as *Those*

Amazing Animals, Real People, and *60 Minutes* all had their own versions of the issue and each had its own special focus on cruelty or the extinction theme to bring out as an exposé of the hunt.

Although millions of people became aware of the fact that a slaughter of white seal pups by club and knife was going on every year, and that this was protested on an annual basis, no other information about the hunt was made available on a wide scale. There was no televised discussion of the mechanics of the hunt, the economic atmosphere in which it was conducted, nor any biological information about the seal herds or their numbers.

No other information being offered, it was widely assumed that there was no moral or humane or economic justification for this slaughter, and that since it was indefensible, it should be stopped.

For over a decade, various animal welfare groups appeared to be taking turns going to the ice and filming their members in the act of protesting the seal hunt by crouching protectively over seals, or dying their coats to ruin them for market, or physically blocking sealers' access to the herds with their boats or their bodies.

Greenpeace, Cleveland Amory's Fund For Animals, and Brian Davies' International Fund For Animal Welfare all conducted early protests. Later on, Humane Society of the United States and Animal Protection Institute, recognizing an opportunity to increase their public profile as humane defenders of the seals, participated in a smaller way in the Gulf of St. Lawrence and off Newfoundland's coast.

All these organizations and some others which never went to the ice or actually held an active demonstration anywhere, began to use the seal hunt as the most successful sort of fund raiser. Each group gathered to itself photographs of the hunt and the support of the famous to publically denounce the slaughter and to claim that, with enough funding, something could be done to stop it.

Each group advocated letter-writing campaigns to Canadian and United States officials, and some declared that boycotts of Canadian goods and self-imposed bans on travel by tourists should be undertaken in order to demonstrate outrage.

However, the annual hunting of harp and hooded seal pups by Canadian fishermen and Norwegian sealers continued without interruption. Tourism in Canada was never measurably affected, nor was the sale of Canadian goods.

Cleveland Amory advocated a boycott of Canadian resorts and general travel in Canada. However, the exchange rate became increasingly favourable and Americans were not inclined to carry humanitarianism that far.

In the early and mid-'70s the movement gained impetus due to the interest in "environmentalism," which stressed an increased awareness of dangers to the earth and increased fears related to new information on wildlife extinctions. The anti-hunting movement gained strength as a part of this, and animal welfare groups used the extinction theme in bolstering their stand against all sport hunting and in protest against the inadvertent killing of dolphins in commercial fishing nets.

In 1972 the United States passed the Marine Mammal Protection Act, which forbids the importation of the by-products of any marine mammal into that country. This was originally intended to be of immediate benefit to dolphins and encouraged the development of nets which would allow them to escape unharmed. However, there had also been a growing interest in whales and seals, and these were included in the all-encompassing MMPA, due to both fears of extinction and to claims of inhumane killing methods for all.

The Marine Mammal Protection Act does not cover those fur seals which are killed in Alaska's Pribilof Islands. These animals are culled yearly, and bachelor males of a certain length are driven to an isolated spot and clubbed, bled and skinned for the fur trade. A Congressional investigation into the response of the population to this controlled hunt found that the seals had actually increased in numbers during times when they were hunted. Furs from *this* hunt were not prohibited from import into the United States.

Many people have heard the message of protest, have learned about seals only from the literature of the protest industry, and have believed them to be highly intelligent, loving parents, mysterious biologically, and in extreme danger of extinction should the hunt continue. They have accepted the message of protest: that seals are intrinsically so valuable and so sensitive and so intelligent that to *ever* hunt them would be an inhuman, cruel, wasteful act.

Brian Davies, executive director of the IFAW (International Fund For Animal Welfare), has even claimed that seals are dog-like in their personalities, habits and intelligence. Davies has further claimed that seals have a human characteristic: they cry copious tears when frightened, distressed, or deprived of their newborn infants. This weeping phenomenon has been described again and again by him in his recollections of harp seal dams returning to the slaughter scene to search for their young and finding nothing but bodies; stiff, bloody carcasses, "wasted" for their fur.

Davies' listeners and followers are often pet owners who have been raised on Disney movies. The more recent of these emphasized wildlife as personalities. Instead of animated cartoons, the newer and more sophisticated Disney films have been cleverly pieced wildlife scenarios. Each has featured a particular animal or its "family" in an annual cycle or a particularly important aspect of its life. Bobcats, mountain lions, sea otters, beaver, raccoons, skunks, bears and birds have all been portrayed as clever, amusing individuals. Each film has been narrated by a deep, fatherly-sounding male voice which has a bemused quality. Although these films may have been produced with the best of intentions, to both entertain and educate the public about our wildlife, the effect has been to distort basic realities.

The primary misunderstanding has been about the role which predation has always played in a balanced environment and the fact that man is as much a natural predator as any other creature. The modern protest against Canada's seal hunt has billed man as a supreme villain, apart from the natural order and interfering in marine life cycles in an entirely adverse manner. This message has been repeated by all manner of celebrities, from Jacques Cousteau to Bridgette Bardot, from the late Congressman Leo Ryan to Mary Tyler Moore, Loretta Switt and Ed Asner. Each has cried for the seals or spoken against the hunt as brutal, cruel, unnecessary in this age of new synthetics, and economically unjustifiable in the areas where the hunters live.

Much of this message has been purveyed by respected newspapers, nationally-televised network news programmes, paid advertisements on television, and the print media. The context has been international news. The protesters have been labeled "environmentalists" and "ecologists." Their message has not been questioned because of the tradition of credibility in which it has come to maturity. Network news programmes and documentaries have reported that Greenpeace, The Fund For Animal Welfare, IFAW, the Humane Society of the United States, the Seal Rescue Fund and The Animal Protection Institute have protested the killing of "endangered harp seal pups" on the "nursery" ice in Canadian waters. The "cruelty inherent" in the butchering method (clubbing) and the allegation that many animals are actually skinned while still alive, add to the story. Shock and horror on the part of the audience are assured. Canada and the sealers are given no voice to counter the above bad news. The complex of information about the hunt grows to include fears that the seals are in imminent danger of becoming extinct due to the hunt's excesses. Western civilization cannot help but be against the seal hunt. It has been defined as bad with no qualifiers.

Since conservation measures were first enforced in 1965, the harp seal population in the western Atlantic has apparently grown steadily despite the yearly harvest of animals, eighty percent of which were non-breeding young of the year. However, no mention of the growth in population or general seal health improvement has been made by protest groups. Thus, the general public has not been aware of this, and has had to rely on the only figures given. These figures have not reflected the reality as reported by professional marine biologists, and have been, in many cases, entirely unverifiable.

In the late '70s and early '80s another bugaboo was added to anti-seal hunt lore; the threat of loss of genetic diversity in harp seals due to the great depredations alleged by environmentalist groups to have occurred in the 1950s, before hunt quotas were taken seriously. The Washington, D.C.-based Center for Environmental Education's Seal Rescue Fund has used this argument in support of its "save the seal" donation drive. Loss of genetic diversity within the herd and loss of diversity in the general environment due to extinctions have been vaguely alluded to in Fund literature as a danger which we must not ignore. The message has been that hunting has caused this loss and that all hunting must be stopped pending further study. The Seal Rescue Fund does not, however, claim to be conducting or funding such study. No reference is made in its literature to Canada's programme of marine research, and a member of the general public would not be aware that such a research programme has ever been in effect. The Seal Rescue Fund would seem to be not entirely objective in its attempts to realize its stated goal, to inform people of the status of marine biology in this area.

Nearly all Americans, most Europeans and British citizens, much of the Australian population, and many Canadians believe that Canada has allowed a great crime against nature in continuing the hunt for harp seals. The fact that most of the animals taken are "babies" or pups in the whitecoat stage has been the most objectionable and the most protested element. Colourful portraits of seals "grieving" over their slaughtered young exist in profusion. The following is an excerpt from Brian Davies' February 1982 newsletter to members and supporters of the IFAW:

> The poor souls I am writing about this time are seals — animals that have a level of intelligence and sensitivity that is at least as high as that of my own loveable black labrador dog, Happy.
>
> For centuries, defenseless harp seals and hooded seals (mostly babies) have been subjected to the most *horrifying torture* every spring off the east coast of Canada.
>
> Cruelty as dreadful as anything I have seen in my twenty-one years of working for animal welfare — as they are massacred

for the luxury fur and leather industry — *or sometimes just to satisfy someone's dark lust to kill.*

This is what happens at the seal nursery on the floating sea ice in Canada's Gulf of St. Lawrence and off the coast of the Province of Newfoundland, where the females have their pups in February and March of each year.

Frightened baby seals of some ten days of age are separated from their mothers and (crying pitifully) are then brutally beaten over the head and sometimes the throat with clubs or murderous ice pick-like weapons.

Bleeding from the nose and mouth, the baby seal is then rolled on its back and the skin is violently ripped from its still trembling body.

The monstrous truth (and I have personally seen it happen) is that some baby seals suffer the dreadful pain of being skinned alive.

After the killers leave the ice nursery, the desperate mother seals return to the shattered carcasses of their babies.

And I have personally shared their loss, sitting beside them through long, lonely hours on the ice-pack as great tears flowed from their grieving eyes.

And I have not forgiven, nor will I ever forget, those who are responsible for these terrible things.

Millions of people have responded to Davies' messages and have sent their donations to the International Fund For Animal Welfare. The millions of dollars have been spent on publicity, paid advertisements against the hunt, and the dissemination of erroneous information about seals and their numbers. Davies' "information" is entirely disproved by the findings of all professional marine scientists who have ever conducted basic research on seals and their habitat. The following story is based upon genuine scientific research into the behaviour and biology of the harp seal. It will serve as a start in the process of re-examination which is necessary to an understanding of the realities of the Seal Wars.

II

Seal Time

The old seal glides from the depths to the ocean's dark, calm surface in one fluid movement. Her black nostrils open and flare and her lungs balloon full, aching for oxygen after twenty minutes under the sea. Gradually, her system adapts to breathing air again. The great reservoir of her blood finds valves opening as she breathes and her circulation reverts to that of her ancestors, who had never gone under the sea for long periods.

Now, all her organs receive newly-oxygenated blood as she breathes air once more and they begin to function again, as if there had been no pause. The old seal is not breathing hard. It has been a routine dive and, slipping effortlessly through the swells to the nearest ice floe, she clambers up for a rest.

As she hauls her three hundred pounds out onto the snowy ice field, a stiff wind begins to dry her sleek black and white coat. The short, bristly hairs shed the salt water easily, and her body shines in the waning afternoon sun. She closes her eyes against the wind, as even her constant copious, syrupy thick tears seem unable to shield her from its sting. She dozes comfortably.

It is late February in the Gulf of St. Lawrence. This old seal and her peers have gathered since December by the several hundred thousands to gorge on the caplin. These small, smelt-like fish are the major energy supply which will be used to build up reserves needed for the nursing period yet to come. Each old harp dam feeds heavily, and the herd consumes thousands of tons of fish.

Now, all are rolling fat. The blubber is white as hog lard, four inches thick under the skin, and serves as more than enough insulation against the constant freezing cold. These seals are not overweight, however. This structure in the seals is protection against the cold, a buoyancy mechanism, and a last resort against energy loss if fish become scarce or when feeding stops. The meat below the fat layer is rich and red, and quite lean lately, as each female nourishes the pup within her. Each fetus is growing rapidly and its movement can be felt. The time for birth, seal time, is only a week away.

As the old harp dam rests quietly on the ice, she tolerates the subzero cold with no signs of discomfort. She does not shiver or try to curl up in a ball. Occasionally, she shifts position a little as the sun comes out or the wind lets up in order to cool or warm herself slightly as she feels the need. Her blubber layer, however, keeps her temperature constant, blanketed securely. Although other females lounge nearby, she and they seem to ignore each other's presence. It is not yet time to feel territorial about a patch of whelping ice.

Only underwater, while they fish, do any seals bother to communicate verbally. Most of the callers are male, announcing the start of the breeding season. Then their songs, made in the throat, come with deep rumblings, odd bubbles of sound that vary with pitch and duration as seals rise and fall with the currents.

Most females are still feeding and will not stop until the pups are born. Brief rest periods are interrupted with long dives, and long hours in the icy water. The caplin are there for the taking. Their vast schools glisten and hover in the dim green light which filters down two hundred feet below breaks in the drifting ice.

The seals, their dark shapes coming from out of nowhere in the murk, suddenly materialize to turn and glide in relentless unison with the fish. The number which are swallowed are as nothing to the school, and when the feeding is over for the time being, the hunters rise languidly, their bodies slowly breaking the surface in the stillness, as if there were no urgency that they breathe air once again.

The seals all feed together, their bodies herding the fish until all are satisfied. These are not lone hunters, but social creatures, working in concert not because it is necessary, but because it is pleasant and, nearing the breeding season, because there are so many seals that to remain alone would be an impossibility.

When all are tired and full of fish, they rise towards the light. Heads pop up everywhere, and their dark sleekness dots the swells in silence. Some slip down for more exercise and more gorging pleasure; others ride the gently-heaving surface which is more home to them than the ice, and requires less energy from their bodies as it presents a constant temperature and shelter from the winds.

By the first week of March, a change comes over the herd. It is seal time, and for the females feeding will stop completely until their pups are weaned. As the time for whelping draws near, the waters become choppy with the increased spring storms, and more and thicker ice, driven by these constant winds, comes bumping and grinding, pan against pan. The sea ice collects in vast, solid, snow-covered fields. Massive ridges, built up by the pressure of wind and current, cause occasionally impassable barriers. House-size chunks of green-tinted, yards-thick, rock-hard ice heave up and become a solid, jagged, ever-changing obstacle to any creatures which move among them.

Yet some areas are as smooth as a parking lot, unbroken expanses which might have been fields. An occasional lead of open water will first appear as a mere crack, then widen to a river-size opening, winding through the ice as though it had always been there. The icescape is forever changing, both above and below. Yet the seals cope with this lack of familiar landmarks, moving with the ice as it drifts.

When her birthing contractions begin, the old seal comes out of her watery element to find her own area to have her pup. This will be her territory, a small patch in which her newborn will stay for its first few days, and into which no other adult will come while she is there.

A few other females within a hundred yard radius have already whelped, and their youngsters shiver in the cold wind. Although it is March, there is no warming trend in the weather.

The dam's eyes water freely as she surveys her surroundings. The tears pour down either side of her blunt muzzle in response to the cold dry air. Her eyeballs bulge, somewhat bulldog-like, giving her a rather nearsighted appearance. All these seals seem to be continually weeping, but it is an entirely natural process, unaffected by events around them and no more indicative of their psychological state than is sweat on a horse in the sun.

Nearby is a crater-like hole in the ice, kept open by many seals as they constantly dive from and return to the shelf. It is used so often, by so many, that its rounded edges form a crater big enough to swallow a man. The edges are smoothed and raised, continually being built up by more ice, then worn down again by the heavy soft bellies which slip through it.

In areas where leads are few or far away, these breathing and exit holes number in the thousands. Seen from the air, they give the scene a moonscape quality, except for the obvious trails in the snow, coming together like the spokes of a wheel, all leading from the crater to each dam's whelping area.

This old seal claimed her patch quite near the dive hole, and the comings and goings of those around her always cause her to raise her head and the upper part of her body in a threatening posture, the "alarm" signal for the harp seal. This nearly vertical position, reared back, erect, with wide open mouth whispering a snarl, is assumed only when she feels too closely approached. It clearly means that no other being should come too near, as she might lunge viciously towards it. Consequently, adult seals on the whelping ice avoid each other's territory as much as possible.

The birth is announced by a gush of fluid which steams on the ice, now swept bare of snow by the dam's circular motion, as she drags her body with the fore flippers, claws digging for a purchase as she strains. There has been no other sign that she is ready to deliver. The pup slides out headfirst, a yellowish, flat looking wad, wriggling slightly from side to side. The membranes hang loosely on his flanks for a few minutes, then slide off into the snow at the edge of the patch. His head lifts and he moves weakly toward the sensation of warmth, the huge presence of his mother. She nuzzles him somewhat, sniffing his face and listening to his mewing cries. The combination of scent and cry, peculiar to him alone, is imprinted on her brain and she will not forget that this one is hers.

It is nearly an hour before his black mouth finds the two teats which she thrusts out toward him, everting through her thick hide in the lower abdomen. As the cold of the surrounding air cools his birth heat, he finally begins to shiver. This shivering response, plus the hot thick mother's milk, plus a heat exchange process in brown fat under his neck skin, helps his body temperature to rise to a normal level. He shivers for three hours, and the heat thus generated dries his lanugo, the yellowish birth coat.

His mother never really cuddles him close, or tries to shield him, although her interest in him is constant. Thanks to good weather, no freezing rain, sleet or snow, the pup warms up and fills out a little with his first meal. The critical stage in his life has been passed; this pup will live.

For his first three days of life, the pup nurses whenever the dam comes near from her dives with the others. The sun comes out after the storms of the week before, and although the wind is fierce and unrelenting, the pup sometimes feels overheated. His blubber layer has developed around him and serves as a body blanket, covering all but his flippers. When he feels too warm he holds the fore flippers out to the cold air, away from his sides, almost as if he were balancing on a beam. This posture helps him to feel better, since the cold air cools the blood which comes to the surface there, and this cooler blood is carried down into his core.

On sunny days he does this quite often, or drags himself to the shade of an ice ridge where he can cool off. Even though he isn't very intelligent he learns what to do to be most comfortable in his surroundings.

Only the gulls share this environment with the seals. They come wheeling over the herd, heads tipping, as their sharp eyes find the welcome signs of each new birth. With glad screams they descend and fight with each other for the fresh placentas and the hot red membranes. Bird footprints and the drag marks of each struggle for this rich resource are left to freeze along with pools of blood and seal urine and liquid yellow seal feces, until snow drifts over the patch again.

Seals do not avoid any mess around them; it is not a condition of which they are aware. Adults and pups both drag themselves through gore and filth and pay no attention to it. Most messes quickly freeze and are covered with drifting snow. The entire environment gives the impression of clean, wide, untouched expanses. Seals always look clean, although there is no grooming of pups by mothers, or of adults by each other. Pups are seldom stained for long, as they move in newly-blown snow. There is never any odor strong enough to be noticed by humans and there are, of course, no flies or other insects in this arctic place. During his first week of life, the pup becomes more and more active, dragging himself around his whelping patch and wailing for his mother while she is gone. Unlike the adults on the ice, which hiss and emit breathy snarls when disturbed, the pup vocalizes his cries. Harp pups' voices have a hollow, distressed sound, high pitched and wailing. They sound much like the gulls which occasionally plague the youngsters if they blunder onto a placenta-raiding party. Other seals have come nearby to whelp, and finally, twenty-five adults are using the same hole in the ice, coming and going all day. Those females most recently fresh, stay with their young the longest time periods, cementing the bonds which last for only nine full days of lactation.

Mother seals do not stay near their young to protect them, as this has not evolved as a necessary survival pattern. Their function is to provide the milk for only as long as necessary. This fluid changes greatly during the nine days it is available to the pup; at first its fat content is about twenty-five percent and it is most plentiful in volume for the first few days. As the pup matures, the fat content rises and the water content drops until the stuff becomes forty percent fat and very thick and greasy in consistency. The mother seal spends less and less time with the pup during this change, so that his weight gain, although dramatic, is not dependent on frequent nursing but on the increased energy content of what he gains in only two feedings a day. The old seal's youngster gains weight so fast he seems to swell by the hour. His birth weight of fifteen pounds increases so rapidly that by his tenth day of life he weighs eighty-five pounds and is nearly as round as he is long. Most of his weight is blubber. This insulates him from the cold and, together with the inner body fat gained from nursing, will see him through the time when he is abandoned and before he starts to feed on his own.

A less fortunate pup, born near by, will be dead by its tenth day. It was two days old, and had nursed ten times, when its dam failed to come back from an underwater trip. This starveling's cries grew weaker as it constantly searched each returning female. But none would have her, this one which had ceased growth. She looked lean and flat in comparison to her peers. Her eyes, leaking a constant wash of fluid, looked huge in her long face, and her hair appeared longer than did that of the others. It was because her skin was so loose. The others' bodies had filled out like stuffed sausages in comparison to hers and her pelt began to hang and sag on her now-bony frame. Each female greeted her with rejection; harsh shoves aside, nips on the nose, but never the opportunity to nurse.

The orphan began to feel the cold by her sixth day as her meager blubber supply was gone. She became less and less active, and by her eighth day ceased to strive in her futile search for the mother who would never return. She only mewed weakly when she perceived movement around her, and as her internal temperature dropped she finally lost consciousness and froze as she slept, a sad little rag of natural mortality. Snow drifted over the cold form, and stayed.

The dam never returned from her last dive. Perhaps she had stayed below too long, and drifting ice had caused her to lose track of a lead or a breathing hole. Or maybe she had come up in a narrow lead just as it closed, millions of tons of ice crushing inexorably together, throwing up a ridge of sharp broken edges, her body pulped between as if it were nothing. Perhaps she had simply and suddenly succumbed of heart failure. Whatever her fate, there was no hope for her desperate and dying offspring.

The two of them meant nothing to the herd. No other creature would miss them, nor would they be of any particular use to the rest of the life systems around them. Their bodies, infinitesimal in the ocean vastness, would eventually drift down and disintegrate. They would become one with the other nutrients which feed the plankton and the krill.

But the fourth pup to be born in this patch, the youngster belonging to the old dam, continues to thrive lustily, along with most of his peers. These pups have all been born within three days of each other. They all look alike, and all react similarly to the coming and going of their mothers. The dams spend longer periods away from them each day, either lounging on the ice by themselves, swimming in the open leads, or diving together beneath the surface. Overhead, in sunshine and at night, the parasitic pups' cries become a mournful, frenzied din as they lunge clumsily around on the shelf, alternately wailing and resting in the snow. They do not develop play patterns of interaction, one with the other, but occasionally have a scratching tiff with razor sharp front flipper claws as they grow heavier and more active.

Helicopters frequently whirl overhead, as federal fisheries officers and a group of marine biologists make their way north to the main herd where tagging and counting are taking place and where the annual pup slaughter in the Gulf is about to begin. The men are impressed by the size of this patch of seals which could not be reached by the landsmen, frustrated by miles of impassable ice, so this part of the Gulf herd never encountered men. Seals are counted from the air photographically; their bodies are recorded with the use of a heat-sensitive film. White pup bodies radiate energy they have collected from the sun and their dam's milk, and their white shapes, invisible on regular film, show up black on the ice. Photographs are superimposed on a grid, and thus their numbers can be counted.

The location of each patch of seals noticed from the air is relayed by radio to recorded telephone messages at a data collection center in the Magdalens. The landsmen can call the number and find out if the seal ice can be reached through open leads by their wooden fishing boats. However, the ice is always moving. The coordinates of a concentration of seals noted in the morning will be entirely different by the middle of the day. The news of resource location is regularly updated, as government employees happen to record information while engaged in research and other duties. No missions are flown specifically to collect information for seal hunters, sometimes to their disgust. Since the ice moves as much as thirty miles in a day, access to the herds is constantly shifting. A man's windfall opportunity can easily and suddenly turn into disaster; his boat might be sheared off at the water line, stranding him and his crew until they are missed and relocated by another passing aircraft.

Near the northern end of this great herd of seals, currents in the Gulf cause open water leads to widen and join one another just in time for the annual slaughter. Many fifty-foot boats, and a few large vessels, come out from the islands prepared to spend a week or more in the spring quest for fat, meat and fur. This is serious business and hard, risky work.

Seal time has been anticipated for months. Money has run out for each household and the stores of pork and saltfish are nearly gone. Everyone wants the chance to earn some hard cash for his pelts and to bring in some fresh meat. It is a time of great excitement, and hope for economic relief. There have been unpaid bills, and nothing to do all winter long. The fish, man's only other resource here, have been safe under the ice which makes net fishing impossible. As soon as the spring season for whitecoats opens, men take their supplies and head out for the seals. When the herd is reached, each longliner is moored to a great raft of ice and the hunters, wearing their oldest and warmest clothes and spiked boots, set out on foot to select their prey.

Pups here are in prime coat, and carry a fine load of fat. Most are six to eight days old, and just right for market. Most of the dams are gone for long periods of time, or will leave if the men come into view. It is a rare young dam, whelping for the first time, that will stay by her pup and act aggressively toward man. Prior to 1965 most of these would be routinely killed and processed along with the whitecoats. But since that time, regulations began to change as seal management plans matured. The females were recognized as a mainstay of the herd and rules about how many of them could be taken changed with each passing year. By the late '70s and early '80s regulations resulted in increasingly smaller quotas of adults in the allowable catch, and by 1982 none could be taken in the whelping areas.

By this year, therefore, those seals which stay on the ice with their pups are passed by, and these pairs are not bothered. The pups taken are those left alone on the ice. Since there is no shortage of seals, this is not a regulation which causes any hardship and men abide by it. Besides, the fisheries officers are always at hand and have the power to revoke one's license if regulations are violated.

The herd is so dense here that the crew of the *Techno Venture* collects the large vessel quota in three days, and the longliners each become so full that they ride low in the water. Men are tired and hurry to finish the job before the next promised storm can blow in and trap them in the ice.

From the air 150,000 white coated pups can still be seen, covering the ice to the horizon. Their numbers have been depleted only within a few square miles of this concentration, where men could walk from boats, jumping from pan to pan, "copying" their way along.

The scene near the boats is nearly quiet now; only a few pup voices are still to be heard, and red trails lead from each red blot where a pup has been clubbed, bled and skinned, to a large, fainter red smear where a pile of pelts has lain. Fainter red trails and footpaths in the snow lead to the ice edge where pelts were winched on board. Small red carcasses, most the size of beagle dogs, lie freezing everywhere. They are not pretty; front flippers have been taken off for the rich red meat, and occasionally, heart, liver and kidneys have also been removed.

The rest of each carcass is not worth much at this lifestage. Remaining flesh is thin and seems jelly-like in its softness. There has been no time for any muscle development in these pups, except for the front flippers which do all the work of moving them about.

Each skull has been crushed by the club blows and bears testimony to the efficiency of the method. Instant unconsciousness has resulted from the first blow, and has been made irreversible by the mandatory next two. Brain destruction is followed by the next step: each pup is stuck with the knife in the chest, between the flippers, and the major arteries are severed. Death within fifteen seconds is assured, as the animal quickly empties out all of its life blood. Dead bodies twitch, and most thrash side to side in a swimming motion, as they bleed out. The compression of the spinal cord as the first blow is struck causes this reflex. It is identical to that seen in decapitated chickens, stunned hogs, or beef in a commercial slaughterhouse. For the sealer, the reflex action means he has done a good job. The pup is not conscious. To anyone who might have dropped in to "observe" the hunt, the scene would have been one of complete shock and horror because movement to the lay person would be a sign of life, with the possibility of pain and some consciousness. But the pup is dead.

For this year, slaughter in the Gulf is virtually over for whitecoats. This kill scene will freeze and become snow covered. The ice will drift off, break up, and melt, and the remains of the seals will feed the fish and the gulls.

While the first group of men has found their seals, done their work and are leaving, another is hurrying to the scene. The Greenpeace ship *Rainbow Warrior* has been having terrible luck this year. Her captain has misjudged ice conditions in the Gulf, and found his ship locked in tight. It is only with the greatest of effort, with chain saws and some dynamite, that the old vessel has been freed and has finally found her way north to the slaughter scene.

Now the annual media event, the spraying of the seals, can take place. It has to be accomplished quickly before the weather breaks and high winds and blowing snow ruin the chance for decent videotape.

Government icebreakers come within a mile of the *Rainbow Warrior* as she jams up alongside pack ice at dawn, on the day *after* the *Techno Venture* and most of the longliners have loaded their hulls to either quota or capacity. Government helicopters hover off the *John A. MacDonald*, waiting for the first of the protesters to hit the ice and head for the seals.

Greenpeace volunteers from several nations suit up in their survival gear, complete with arctic insulation, blaze orange, studded boots and heavy pressurized tanks of green vegetable dye in a water solution. The elaborate and well-financed expedition is ready to begin.

Thus equipped, and with video equipment operators in attendance, green lights on cameras are noted and the signal is given. The volunteers clamber down onto the ice, wondering how they will be able to maneuver quickly enough to reach the seals, apply the dye, and manage to elude the arresting officers while bundled up so tightly they can hardly move.

A few pups are located and cheers go up as cameras record the first liberal spray applications. Those pups covered shake their heads as the green stuff runs down from their ear holes and splatters onto the ice. They are covered with a few passes of the gun nozzle from head to tail down the centre of the back. The green wetness soaks through the fluff of the birth coat, which is not water repellent and which thus loses its insulating qualities until it dries. Later, the dark colour will serve to block the sun from penetrating in the usual way through crystal white hair to the skin. Some ability to use solar heat is thus lost by the few individuals which have been sprayed. Their whiteness has served, not primarily as camouflage, but as a part of the intricate system which has been evolved to take advantage of the sun.

No pup thus sprayed will die as a direct result of the artificial colour, yet a net loss of energy is the result for each animal. Warmth has to come entirely from its inner reserves, rather than from the renewal derived from sunshine each day.

But the mission of mercy continues. As cameras buzz, more and more pups are found and covered. Finally, carcasses from the day before come into view. Blood spots on the ice are recorded, as is an adult harp which has come out near a carcass, some eighty feet away. At that distance she becomes alerted to the men and assumes the erect, high head posture of a seal on the defensive. As a man slowly approaches, she turns to face him with mouth wide open, fangs gleaming, myopic eyes bulging wide and pouring their fluid. At her chest, under her flippers, is the body of a pup which she has happened to come across, and which she has paused to sniff and nuzzle. She has nosed it only briefly, as it is frozen, but the circumstance of her position over it, with the tears and the aggressive, seemingly protective stance, make a striking tableau. (In the past, more than one group had made use of such a scene in its photographic depiction of the "tragedy" of the hunt; such pictures were used to portray a typical example of a grieving mother returning after the slaughter to search for her dead pup.) As the men come as close as fifty feet, she turns and slips down the dive hole.

Nearly eighty pups have been found by the wide-ranging Greenpeace and covered with green dye. Men near the slaughter area are arrested first, and handcuffed for the cameras. The others, slipping and stumbling off in the other direction, are spotted by helicopter and rounded up one by one. All will be taken off to court on the Gaspé Peninsula, judicial headquarters for this part of Quebec, and duly processed. The charge is violation of the Seal Protection Regulations. Sections are quoted pertaining to the illegal frightening of a seal herd, and approaching within one-half mile of the seals. Protesters are fined, and fines are immediately paid by Greenpeace.

The Greenpeace mission has again been a success. Cameras have managed to record the current slaughter scene, the volunteers dying seals, and the arrests. The old *Rainbow Warrior* is seen silhouetted against the dawn sky of the Gulf and the cries of seal pups have been recorded.

When the clip appears on television news programmes in the States and elsewhere, the message will be plain: Greenpeace is still trying to save seals by covering them with symbolic green, ruining the pelts for the hide market. The green ones will be spared a horrible death. Greenpeace volunteers gladly suffer the great danger of the ice mission, and the great indignity of the arrest process, for the greater cause; public awareness has again been raised to the realities of a crime against nature. Viewers will appreciate the irony of being arrested for trying to prevent the killing of baby seals.

One fact, however, is not included in news coverage of this event: the seals had been sprayed *after* the *Techno Venture* had reached her quota and sailed away. Her men never had to choose between a green seal and a white one.

Two harvests have been completed. One realized pelts, fat to be rendered, and fresh meat. The other is much more profitable. Film produced by Greenpeace will bring in thousands of dollars a day, from March through about October, when the annual money flow finally ebbs. Donations from animal lovers will be added to the tax exempt revenue of the organization and will surely finance another mission next spring. Since the film clip has been seen on national news broadcasts, a perfect credibility for Greenpeace is assured. The resource of the spring ice is, indeed, invaluable and renewable, as long as the seal hunt should continue.

Only a small percentage of the herd is affected in any way by either the hunting or the protest activity. Fifteen miles to the southwest the old harp and her peers continue life unaware of any disturbances. She has seen both hunters and seal savers in her long life, and remembers none of it.

This year, her pup and his fellows will be out of reach. For them, this year is one of choked up ice fields and treacherous, narrowing leads. In this place, nothing will interrupt the short segment of life when pups demand, and dams give up, the milk which works such miracles of stupendous growth.

By the end of the eighth day of his life, his milk growth all but completed, the pup is an eighty pound lump of complacency. His dam has nearly ceased paying attention to him. In the late afternoon she lunges over to his patch of ice, now empty, and suns herself. It has been a social morning for her as she has dived and cavorted with four hundred others, both males and females, in a vigorous pre-courtship exuberance. The seals spend their time calling to each other under water with hearty, rumbling throat sounds. Again and again, each rises and dives, interweaving with the shafts of light which filter down in pillars from leads and breathing holes. As each feels the need, it rises gracefully to the top, breaks the surface only as long as necessary, then dives again.

Finally, the old seal becomes tired and returns for a time of rest, and to nurse the pup. He has been only a short distance away, sunning himself with flippers tucked beneath his chest, his eyes half shut, the warmth soaking deep into the blubber which is a full three inches of creamy thickness beneath his prime pelt. He is forty-one inches from head to tail, and thick as a log.

The pup stirs as a breeze tells him his dam has returned. His eyes, brimming wetness as usual, widen. He quickly scrambles toward his birth patch near the breathing hole, now a huge, rounded crator worn smooth by many heavy bodies.

As he nears his dam she surveys him with the usual hostile suspicion, and sniffs his face thoroughly before rolling over on her side and exposing her two black teats for the last time. There is not much milk left in either, and it is very thick and greasy. He quickly drains all there is and continues to bunt and pull on her until she flops back down on her underside, causing him to struggle out from underneath with some difficulty.

In spite of his greedy, protesting wails and constant nosing about, she stays firmly on her belly, with teats inverted back into the depths of her body. He has had all the nourishment she will ever provide. Within twelve hours he will be exactly nine days old. She ignores his presence, tucks her fore flippers under herself, and goes to sleep in the waning sun. The pup, discouraged at last, does likewise a short distance away. He is now as good as abandoned, although he has no hint of this, and his mother has given it no thought. She is incapable of any thought for tomorrow, or yesterday.

Nearby, a male harp has hauled himself up onto the ice, after cautiously surveying the seals around him while he trod water in the breathing hole. He bears scars of many lost battles on his dark, sooty hide. The most recent still hurts his left shoulder where an older and heavier male had cut him with a vicious rake of the fore flipper claws. His skin was left ragged in a long, ulcerating slash. This male has never had the chance to breed any females, although he is now four years old. Others have always driven him off, and the mating has been done by the old dog seals. But there are no other males up here on the whelping ice.

Now, his black nostrils flare with the sweet scent of bitches nearing estrus. He moves cautiously, head high, near the sleeping dam and her pup. Her senses of smell and hearing awaken her and she lunges in a rage at the stranger, her bearlike mouth wide open, sharp yellow fangs exposed, her throaty cries a harsh, nearly vocal, coughing wheeze.

Even though she is nearing the time to mate, this is not the place. Instinct still causes her to lunge between the pup and the intruder. The young dog perceives his error and hastily turns, scrambling his bulk towards the exit hole in the ice as her teeth pop inches away from his retreating flanks. A whoosh of water boils up the hole after him, sloshing over the sides. He is gone.

The dam, enraged with the action of two minutes before, is still in motion. She expels her breath and dives. As the water hits her forehead, her system automatically begins to shut down and her body becomes that of an aquatic specialist. Breathing stops. Her heart rate slows from 120 beats a minute to twelve. The massive amount of blood in her body ceases to flow in normal routes. Smoothly and routinely, shunting valves close in her circulatory system. Her heart and brain continue to receive a full flow in a shortened loop of circulation, but the rest of her organs are in a holding pattern of suspended function.

The dam's bulging eyes serve her well under water and she searches for the offender for a full minute before forgetting him completely. He is lost in the crowd, and would have been safe had he come face to face with her. He and her pup and the incident dissolve from her mind as she joins in the familiar exhilaration of the dive with the others, twisting, turning, chasing, and surging ever deeper in great rushing bursts of speed, then floating languidly upwards for a time before swirling down again. All the seals are transformed, when under the surface, to quick shapes whose deep, peculiar mating calls bubble everywhere.

For some it is time to breed. After the dominant males have convinced the others of their strength and determination they pair with those females in full estrus and couple side by side as they swim. For others the courtship has not yet culminated. Old dogs check one bitch after another in a search for those who, with body language and willingness to acknowledge vocal calls, signal the beginnings of full receptivity.

The pup's old mother has become one of the crowd of those seals which are nearing full estrus. Within a few days she will copulate with an old dog with full harp pattern on his coat, and many scars. When they have finished, she will continue to entice others, and perhaps some will be the lucky youngsters of five or so years, mating for the first time.

Although she frequently hauls out onto the ice to rest and sun herself, she never again nurses her offspring. Her interest in him has diminished to nothing with her milk supply. Ten days of lessening concern coincide with her changing hormonal balance. Her interest in finding a mate increases to a frenzy which obliterates all else from her consciousness.

All pups continue to nurse as long as mothers and milk are available. When they find themselves alone, without the familiar routine of returning nourishment, they wail miserably, instinctively using their voices to try and entice a dam's ears. They frenziedly check each returning female and demand the familiar attention. But this is not given, and gradually their mothers spend even more of their resting time in the water, away from the youngsters. The whelping ice gradually drifts away, and the adults do not try to stay with it.

Humans continue to pass over in helicopters, checking on the size of the herd and its position. If landsmen should reach these seals by boat, the officers will land and the slaughter will be supervised. Licenses as well as the men's procedures for killing and pelt removal will be checked. But no man comes near. As the shadow and noise of the craft pass over them the pups seem to swarm and lunge in unison, startled by this presence. When seen from the air, they appear to all flow in a preplanned direction, like a flock of birds evading a harrying predator. All pups are now abandoned, and no coloured adults are seen in the swarm below. Their globular bodies are slightly creamy against the purity of the snow. From the air, the black noses and eyes give them the appearance of a thick brood of maggots, all squirming together.

The cycle of seal time has climaxed again. This year's brood is on its own in an environment to which they are nearly perfectly adapted through their specialization. Yet they are still not complete, nothing like the sleekly cavorting, muscular forms of their parents, the quick and dark sea hunters.

While mammals have no true larval stage of development, as in the insect kingdom, there is still a peculiar, intermediate larval quality to the infant harp seal. The still white pups will not eat anything at all from the end of nursing, at ten days, to the end of their first month of life, or even longer. They are warmed by their energy reserves, by their blubber insulation, and by sunlight. Their pure white hair is soft and wettable, not made for swimming. Unlike the adults, these are not sleek predators, but resting ex-parasites, nymph-like beings only passing time to the next stage of maturity. Not yet to swim and socialize below, for this would disrupt that other stage on which all their world is dependent. Hundreds of thousands of pups, all taking to the water at the time their dams are seeking mates, would confuse the whole balanced procedure of underwater communication and pre-courtship ritual. Down there, nothing is important but the forming of new pair bonds in preparation for next year's seal time. Even eating has been forgotten in this pursuit of the old ones. It would be a wasteful, less productive pastime.

So over the millenia the youngsters have evolved through a passive stage, abandoned on the ice where there are no predators anyway, floating away from the sexual part of the herd, which stays together. The next vital step in ensuring continuation of the species is refertilization.

Most of the pups will live to carry on their kind. The species will continue to renew itself, step by step, as part of the ocean's energy rhythms. As long as the old ones are spared and enough pups live to serve as replacements on a regular basis, there will always be harp seals.

On his eleventh day the abandoned pup, most of his weight now blubber, drags his bulk far from the well-worn area where he was born. He and all the other fat nymphs in this part of the herd expend precious energy by continuing the futile search for their mothers. Their voices raise a hollow din of wailing which goes on desperately, to no avail. By his fifteenth day of life the white fluffy hair begins to loosen and there is a grayish undertinge to his face and back. By the eighteenth day he is a "raggedy jacket." A black stripe of new hair shows through down the center of his back and his face is becoming darker and darker as white hair falls out. All the pups change to dark faced youngsters, shedding out all over without realizing that anything has happened.

The weather again becomes stormy, but they make little effort to find shelter from the wind and driving sleet. They just tuck their front flippers beneath themselves and wait it out, dozing comfortably. The inner stores of fat energy keep them warm and safe, and their new bristly hair coat is water repellent.

The pup dozes his way south and east for two hundred miles, winding through the Gulf north of Prince Edward Island, bumping along with the currents. His ice is worn away as relentlessly as the wind which pushes it through the swells. Each pup is a complacent little slug being borne along a general flow to the east and the Cabot Strait between Newfoundland and Nova Scotia. Soon the open Atlantic will be its first real adventure, and the second most important event since the day it dried off and first drained the dam's thick milk.

On the morning of his thirty-third day of life the pup awakes to a heavy, jarring impact. His ice becomes a crazily tilting, fractured island which grows smaller as the day wears on.

By the next morning the wind is even stronger as the usual war of high and low pressure systems moving off the continent catch him and his peers and toss them without letup. Spray covers them all and some, slipping off into the water, flounder and beat wildly with their flippers as they ride the dark heaving newness for the first time. They are buoyed up so well by the blubber layer that they do not have to work to keep on top, nor do they try to keep their noses above water. They will thrash around for a little while, then dip back in again after a short rest. Soon, the difficult climb back up the ice floe's steep face becomes more bother than it is worth. In order to reach that haven, black claws have to dig furiously, trying for a hold on the slick, steep face which always bobs and heaves. Each in turn, after a struggle, will make its way back up onto the familiar and solid environment of the shelf. But the youngsters, in this one day of new wetness and turmoil, have changed. They become more and more restless and their complacency vanishes forever.

It has happened. Suddenly, these dark new ones are seals! With a vigor and purpose, they go back into the sea, one after the other. As each submerges, its body responds in the deep diving reflex, slowing the heart and changing the blood flow routes. Each looks with amazement and surprisingly clear vision at the underwater world so different from the ice above. Murky greenness and silence surround them and cradle them. They are water babies at last.

The pup expels his breath and dives far down into the green stillness. His body curves and twists and responds as though he has always done this. Ahead, a cloud of shrimp cluster under the ice edge. He propels himself into it, mouth wide open, and gulps in his first solid food. It flows easily down his throat and he gorges lustily on these objects, sucking them into himself in great numbers, now snapping and lunging to follow them until he is full and potbellied with this treasure. He feels good. As the days pass a new strength permeates his entire being. He will never be passive again.

From now on, these *bête de la mer* or beast of the sea stage pups, or "beaters," newly moulted to a dark, sleek coat and newly muscled from swimming, become an integral part of marine life. New energy stores from fish and crustaceans replace their milk fat, and this new layer insulates them when the sun's heat cannot reach them in the water. All are accomplished swimmers, and they head generally north and east, through the open Atlantic. They follow warm currents and the shrinking number of ice floes near which their food supply hovers, and they need no adults for instruction or protection. They are on their own.

This movement is a migration, and although it is gradual and mindless, as it has been for untold ages, it takes them instinctively to the western coast of Greenland where they fish an increasing variety of species and never come out on land. Now the ocean is their life.

It nourishes and shelters them well. Their wastes feed the algae and the plankton in their midst and contribute to the whole as other life cycles enmesh with their own.

More than 500,000 pups have been born this year in the northwestern Atlantic. Those from the Gulf of St. Lawrence and those from the "Front" off Newfoundland all gradually make their way northeast to the same area. All the survivors, that is. Some have died weather-related natural deaths; sleet storms or high winds have taken a certain toll of newborn before they had a chance to dry off and warm up. Some have been early-abandoned starvelings; some have been stillborn.

The humans have taken less than one-third of their number, as usual; approximately 160,000 pups have been slaughtered by men with clubs or guns. Over 300,000, however, have lived through their first few months and perhaps will survive their first season, and more, as seals. Although the hunt quota has allowed as many as 186,000 to be legally taken, the actual take in recent years has nearly always been less. The herd continues to grow steadily, in spite of a yearly harvest of the non-breeding young. More seals are produced than drop from the ranks, the food supply is plentiful, and the cycle is virtually uninterrupted. Some two million seals ride the western Atlantic migration routes. Another million inhabit the Soviet and Norwegian Arctic, migrating to more southerly waters in similar patterns.

It will be four years or more before the new ones mate for the first time. The years between will be spent joining the mature herd in their first fall back in southern waters. From that time on, they will migrate north and south together with the old ones as seasons change and ice and fish move around them in a yearly pattern.

Summers in the high Arctic, gorging on polar cod and crustacea. Winters in the Gulf and off the Front, filling up on cod, herring, mackerel and caplin. For ten months of the year all they have to do is eat, swim, socialize, and eat some more.

Fortunately, this requires no great amount of intelligence. Most of their activity is genetically programmed to happen through changes in hormone levels brought on by different lengths of daylight and the gradual fluctuations of sea temperature as they move from one latitude to another. One might fairly assert that these are fortunate creatures. From a human perspective, they live without stress or complication, and neither worry nor plan.

The old seal had abandoned her pup on the tenth of March. By the fifteenth she had copulated many times, with several individuals.

By the end of the month, their copulation over, most dogs and bitches are heading farther north to haul out on still intact ice fields and rest. It is time to moult. All those harps over the age of a year now come together to complete the annual cycle in a lazy crowd, lying on the ice and waiting for their coats to renew themselves.

Their hides are now worthless to the humans who have recently coveted them and hunted a few of their number for the slick, short hair coats. Now they are shedding. This necessitates a certain body temperature and a certain amount of daylight and dark, with very little eating and underwater activity. Their energy reserves are nearly gone and they need the rest. So they loll about, scratch when it seems pleasant to do so, and do nothing until the hair is completely renewed and they no longer itch and feel lazy.

The old dam still appears sleek and well. Her blubber layer has decreased since her lactation, but is still thick enough to shield her from bitter winds until her new hair comes in. Her new fetus has been conceived but is still free-floating in her uterus. It will not be implanted and nourished for another month. By mid-May the moult is finished. The seals are slick-haired again. They have gradually been moving north, as their ice has carried them south all this time, and it has been necessary to spend longer and longer periods in the water just to keep ahead of this trend. As most recover they begin to feed more and more, and then to move purposefully into the open Atlantic towards the high Arctic for the summer. Two thousand miles of open water heave before them in a vast, rolling expanse. The important prey now will be Arctic cod. As each dam begins to feed voraciously again the depleted energy stores of her liver and other deep organs are renewed. Now, after a delay of nearly two months from the time of its conception, each embryo moves to become firmly attached by its placenta on the uterine wall. This delayed implantation has given the dam the time she needed to regain her own strength, change the old coat for a new, and rest without the drain of the needs of this new parasite. It has existed in a dormant microscopic state until the flood of nutrients comes in as its host begins to eat regularly once more.

Ten months from now, it will be seal time again, as it has been for untold centuries since seals took to the sea and became a part of its rhythms.

III

The Way It Was
1922, 1969, 1979, 1982

In 1922 an American writer boarded the *Terra Nova* at St. John's, Newfoundland and began to describe "the greatest hunt in the world." His book was published in 1924 as *Vikings Of The Ice*. George Allan England's accomplishment in recording the men and culture of Newfoundland has been compared to works by Kipling for its colour, drama and sense of realism. England's photographs and description of all the gory details of the seal hunt serve as a record of the way it was in the days before resource conservation or cruelty to animals were considered of prime interest or importance. England experienced a goodly amount of culture shock during his trip with the sealers, but came away with a deep admiration for their courage, honesty, sense of humour and inner strengths.

Just prior to 1969 the beginnings of modern day seal hunt protest were being seen and heard on both sides of the Atlantic. *Vikings Of The Ice* was suddenly remembered and opportunely reprinted, with a new introduction, as *The Greatest Hunt In The World*, first in hardcover that year then in paperback editions in 1975 and 1981 during the heat of the international publicity about the annual hunt. The first feature-length movie made in Canada is based on this book and is in itself an historic document and a foreshadowing of the myths and mystique of sealing.

England's experience with the hunt took place in a time before quotas when the world still assumed that the seals, put there by God for man, would never diminish in number. Some have claimed that the world population of harp seals numbered some nine million animals before World War I, but these figures are impossible to substantiate. By the 1920s the annual hunt netted some 700,000 animals taken by up to 14,000 men. Young and old seals, breeding stock and immature surplus young, were all killed as they were available, with no worry for tomorrow. Most of the meat was left on the ice; only a small fraction was used to feed the men who procured it. This was before the days of refrigeration technology. The main goal was the sculps, the fat-laden pelts which were piled up in the hold until the grease ran out of them from the sheer weight and heat of the pile, rendered to a rancid mess with an overwhelming and unforgettable aroma.

The end products sought were not fur coats for the idle rich, as is so often claimed today, but a durable leather. The hair of whitecoated seals is not very tight unless they are stillborn "cats." Technology was not entirely able to overcome this, and so pelts of the white pups were used for the hide itself. Older seals in good coat were used for boots and other items with the nature, bristly hair still on. The other and very important use of sealskin was the fat; petroleum products had not yet been developed for wide use and the massive amount of blubbery white fat on each seal was collected to be rendered into a high grade oil used for a wide variety of products from soap to cosmetics, to machine lubrication and cooking oil. Seal oil is still recognized as superior to petroleum-derived oils because of its tolerance to high heats and friction.

Each spring Newfoundland men vied with each other for berths on sealing vessels. Although their extreme poverty cannot be denied, their anticipation of gaining this particular employment had a great deal to do with their love of adventure and a chance to be a part of the legend of "goin' swiling."

They keenly anticipated the excitement of finding the "main patch" of seals, the "young fat," and of killing them all before any other ship should have the opportunity to reach them. Rivalry between ships' crews was intense and a sense of competition pervaded each man's every thought and reason for being out on the seal ice.

The incredible hardship and misery of everyday life on board the *Terra Nova* and other dirty old tubs of sealing vessels, was accepted as no more noteworthy than the air. It probably impressed England because he had never known or imagined such living conditions, while his friends among the crew had never expected anything different.

The fact that England recognized cruelty to seals by some of his shipmates deserves mention. This recognition served to increase his feelings of estrangement from their fellowship, and it and other things which bothered him give us a hint of what each culture consisted of: their's was a tolerance of hardship and misery, tempered by humour and fortitude; his culture expected comfort to prevail and the amenities of life to be present as a matter of course.

Those who look at the seal hunt today see it through standards more closely related to England's than to those of Newfoundlanders, or of any other people whose living is directly taken from soil or sea.

The Greatest Hunt In The World was reprinted in response to protest against this hunt in which fishermen clubbed infant animals to death. It appeared with an introduction by a new editor with a message which gives the impression that sealing had not changed measurably since the 1920s and which dwelt on the reasons for man's interest in, and propensity to, violent acts against men and other animals.

This approach implied that any butchering of seals could fairly be equated with murder and suicide as violent crimes of passion or despair. Some discussion followed which referred to psychological studies and theories about displacement of aggression and hostility.

An attempt was made to explain the existence and techniques of the seal hunt through a brief survey of Newfoundland's history of political and economic repression. The reader is left with the belief that Newfoundlanders are a unique lot, that they more easily accept cruelty and brutality against animals, not because they are inherently cruel people, but because they are taking out their generations of anger and frustration on the helpless seals which have meant so much to them.

Displacement of aggression theories and historical determinism have both been in vogue from time to time as nations have considered each other's behaviour patterns and the reasons for them. Each approach can be seen as one culture's justification for labelling another as "wrong" in its actions. Each of these "explanations from outside" not only fails to explain origins of behaviour patterns of entire communities of people, but also tends to strengthen feelings of apartness and alienation from these people. Thus, the result is not a fuller understanding and tolerance of differences, but the firming up of previously held prejudices and stereotypes. This is the atmosphere in which rumours proliferate, vast exaggerations are easily believed, and myths are born.

It is obvious that some men committed cruel acts against seals while George Allan England watched. Once in a while a seal would be cut into without having been stunned first. In one instance, England quoted Captain Kean as he corrected a man who dragged his animal near to the ship: "John, kill y'r seal — *Don't* sculp 'em alive." Humane concern was not entirely absent from the scene, and definite standards of decent behaviour were upheld by this strong leader.

From a practical perspective, of course, those men were doing a very difficult and tiring job; the harvesting of hundreds of seal pelts each day required a strong back and great stamina. The pelt had to be removed quickly and it had to be usable. A moving, resisting seal would make the job extremely difficult and the pelt would suffer jags and cuts which would render it useless.

In this situation men vied with each other to bring in the most pelts. The role models and leaders were the men who killed and processed the highest number. These experts knew that a stunned, bled seal was one most easily and quickly sculped. No one had time or energy to waste in trying to skin live animals. A dead seal will not try to get away, and it will not bite while it is being worked upon. To ignore this reality and fantasize that skinning alive was a commonplace event, or one which was considered sport, is to ignore all the evidence to the contrary, even in England's account. Adult seals were shot in the head, pups were "batted." The author saw some rare and cruel happenings, and described them because he felt they deserved notice and discussion. They were, at that time, a part of it all.

There has been much said about the "bloodlust" to kill which was a part of the atmosphere as the ship approached the whelping ice. The absolute frenzy with which the men disembarked and set out amongst the whitecoats, and their disregard for avoidance of the blood which soon covered them, was a shock to England. He had never witnessed anything like it. He described men with blood on them from head to toe, smoking their pipes and having their tea and bread with no regard for cleanliness or appearance. He marvelled that a man would wade in warm pelts to waterproof his boots, and left that photograph out of the first edition because he felt it was so shockingly abhorent. He gazed in fascination at a young man who wore seal hearts on his belt and who held one, still pulsing in his hand, not with revulsion, but wonder that it could still beat.

England noted one young man who was wearing a string of seal hearts on his belt, and he photographed him with his "trophies." It should be noted that England had no other frame of reference for this behaviour, and so he labeled it "trophy display" and let it go at that. He did say that it was not something commonly done, and he had no other explanation for it.

This writer mentioned the wearing of trophy hearts to a modern sealer who is familiar with the book. His opinion is that since England lived in the Captain's quarters on that trip and did not eat with the crew he had missed the entire point of seal heart collection. Every sealer knows that the first few meals of seal meat every season are apt to cause severe intestinal distress. Seal meat is very fat, much like pork or roast duckling. A big meal of flippers is no way to start the new year because severe dysentery would be the result. However, a moderate meal or two of seal hearts, which are noticeably lower in fat, is a practical way to break in one's system. Then the body can tolerate flipper or roast ribs without distress.

There is no valid way to discuss or explain satisfactorily the behaviours of another culture without at least a smattering of the tolerance which comes from an anthropological perspective. England recognized his difficulties with this, as he could not help comparing his own world to that of poor Newfoundland. His vision was clouded by his own cultural standards, but improved as he came to know the people as individuals and friends and to know what was so important to them about "goin' swiling."

Those sealers of the *Terra Nova* who worked hardest, produced the most, and obeyed the captain, were rewarded with the admiration of their shipmates, and with fame which lived after them in the memories of their friends and sons. To persevere and succeed at sealing, one had to be tough, impervious to pain and illness, strong, brave on the ice, quick of foot, and willing to work until darkness made more impossible. A successful trip meant that one's ship brought in more seals, and a heavier load of fat, than any other. This gave one a share in the legend which would develop from that wonderful time.

From that perspective, small wonder that after a long and dull journey north to the ice, anticipation would grow until men felt a "bloodlust" on first spotting their prey. Their reputation as a crew depended on the size of their harvest of this resource. Each man's individual reputation and his own sense of worth was dependent upon his performance. No wonder that much of the definition of manhood, in those times and in that culture, depended on a man's proving himself out on the ice. But given the above, perhaps it is more understandable than it would be if we tried to tailor it within a framework of historical determinism.

Those men of Newfoundland were a product of their culture and their times, as are we all. Their expectations of life were different than ours could ever be, for few of us have personal experience of butchering, and killing for any reason is synonymous with violence. Violence is frowned upon, and discouraged in our children, who can never grow up with killers of anything as their role models. Given this, it is understandable how easily we have been persuaded to view as a cruel anomaly the Newfoundland fisherman who hunts seals. We see him as a barbaric, primitive holdover from another time, in a world which doesn't need that any more. The protest movement against the hunting of harp seals has not only encouraged this perspective, but has flourished because of it.

A sports writer for a hunters' magazine once noted that no one seemed to care if a black or brown man was seen hunting with a spear or bow and arrow, or that he killed game with some difficulty over a period of hours or days; yet if a white man was seen in his khaki outfit and pith helmet, riding in his Landrover and killing efficiently with a high powered, scoped rifle, this was decried as "cruel sport." The fact that each had accomplished the taking of surplus game was not appreciated. The one was tolerated because he hunted in a traditional way in order to eat or protect his gardens, and the other was criticized because he did it for enjoyment. The bottom line of relative pain and cruelty to the animal seemed to mean less to the observer than the roles of each hunter. These roles were partly defined as good or necessary, or bad, by the colour of his skin. Inherent racism in the observer coloured the degree of approval of the act of hunting; the white man was "not supposed" to be doing it any more because of his superior advanced state; non-caucasian man was excused because he was "primitive." Perhaps if Newfoundlanders were black or brown they would not have heard so much noise about the barbaric nature of their seal hunt from white Europe and America.

Those who have hired on ships to take whitecoat seals in the spring have not considered that they were committing violent acts, any more than does a butcher in a slaughterhouse. A job was to be done, and the traditional methods of carrying it out were a fact of life. The animal would be dealt a series of fatal blows to the head, then bled out by severing of major arteries, then the skin would be removed, along with its heavy load of fat. The total act takes a Newfoundlander about one minute.

There is little real difference between this and farm butchering, except that, in some areas, there is a tradition of bleeding hogs, sheep and cattle without prior stunning or shooting. This makes the job more difficult, the animal certainly suffers a great deal of stress and pain and dies slowly, staggering around until it drops unconscious from lack of oxygenated blood to the brain. This is a very distressing sight to anyone who is not accustomed to seeing painful death in sentient creatures.

The protest movement has relied a good deal on the imagery of cruelty in Ebbitt Cutler's introduction to *The Greatest Hunt In The World*. He assumes that the book is concerned with violence which must be explained, and which must be remembered in colourful detail. There is no suggestion that the slaughter of seals in any way compares with the slaughter of domestic stock, which is not considered a violent act in the culture of the reader. Apparently no parallels were evident to Cutler and he felt that emphasis on passages about blood and England's abhorence of it would sell to a gentle public.

That Cutler shared England's culture shock is evident as well in his agreement with Joseph R. Smallwood, who had wondered why Newfoundlanders were "unfit for creative and constructive effort." If this isn't blind cultural snobbishness and the epitome of ethocentrism, then one wonders what is. It is ironic that this book, which was an attempt to describe a tradition with fairness and honesty, should provide the basis for the most descriptive and effective protest movement rhetoric. The author acknowledges his own difficulties in coping with unfamiliar conditions and attitudes, and left Newfoundland with both relief and regret. He had never intended his work to be used against those whom he remembered so well, and to whom his book was admiringly dedicated: "To The STRONGEST, HARDIEST AND BRAVEST MEN I HAVE EVER KNOWN, THE SEALERS OF NEWFOUNDLAND."

Fifty-seven years after George Allan England's trip to the hunt, another American, well aware of the controversy which was growing due to newspaper accounts and the use of hunt film on television, asked permission to observe from a large vessel working off the Front. William McCloskey, a writer who already had a working knowledge of fishermen, talked his way onto Captain Puddister's *Gulf Star* in March of 1979. McCloskey lived with sealers as they worked to fill their quota. He went out onto the seal ice, taking his chances with them as they harvested 7,000 animals. McCloskey's main interest seemed to be to investigate claims made by protesters that the operation was extremely cruel for pups and mother seals alike. He had a passing acquaintance with members of Greenpeace, the Fund For Animals, and the International Fund for Animal Welfare, and was well aware of their policy statements and position.

The men of the *Gulf Star* came to trust him as one who would work alongside them, helping to haul pelts over rough ice in arctic conditions. Bill felt that the men behaved no differently in his presence than they would ordinarily, without the eyes of a stranger on them.

He found these Newfoundlanders to be good people, and he saw none of the cruelty described by England in the 1920s. The difference may be attributable to an educational process which comes partly out of a changing culture and partly from governmental direction; by 1965 sealers were being instructed by Fisheries Officers in humane killing methods and all had been made aware that infringements of the ground rules would result in the loss of a man's license. McCloskey saw no seals suffer a painful death. No seals cried out after having been struck *once* on the head, *and no seals were skinned alive*. Although a slaughtering operation is not a pleasant sight, the mere presence of a great deal of blood does not in itself justify widespread condemnation of the operation as a cruel bloodbath. McCloskey was a competent observer. He knew the standards of humane death which are mandatory in slaughterhouses, and felt that the seals were less stressed than is domestic stock.

McCloskey maintained his objective approach by keeping in touch with Greenpeace and with Cleveland Amory, head of the Fund For Animals. When he came to the ice again in 1982, the Greenpeace "kids" were out on the ice, dying seals and hoping to be arrested on camera. McCloskey was able to go from law enforcement planning operations on the Canadian ship *Tupper*, across the ice to where the *Rainbow Warrior* had discharged its volunteers to paint seals green. He stood with them and watched while the government aircraft brought the Mounties over for the arrests on the day after the *Techno Venture* had filled her quota and left for home. His vivid account appeared in *Oceans* in November 1983. This description, and the dry facts of the actual sequence of the departure of the *Techno Venture* and the arrival of the *Rainbow Warrior* which had appeared in French-language newspapers in the Gulf, helped to round out the scenario in this book.

McCloskey's credibility as an objective writer with a humane concern has been supported by two other publications: the *Smithsonian Magazine* (November 1979) and the *Washington Post* (September 25, 1983). In each of these, he has repeated his belief that the seals are not experiencing fright or pain during the hunt.

Both England and McCloskey tried to record the seal hunt "The Way It Was," and each succeeded in doing so, even though the lenses through which they were looking differed in depth perception and in cultural tint. Each man came away from the experience with a great liking and respect for the men of Newfoundland, and with a sense of the great richness of the culture in which sealing and fishing are so integral to the whole. England went to the ice in order to tell the world about the greatest hunt man had ever conducted; McCloskey went in order to determine whether the conditions of that hunt were acceptable in the modern world. He concluded that they were, and has spent a considerable amount of effort since then in defense of sealing as a wise and humane use of the resource in a changing marine environment.

In 1979 McCloskey was unsure of the validity of protest claims that the harp seal was headed for extinction because of the hunt. By 1983, however, he was able to write with confidence that this did not seem to be the case. He based his further defense of the hunt on an October 1982 report by an international panel of scientists. This was the now-famous ICES report, requested jointly by Canada and the European Economic Community.

The International Council for the Exploration of the Sea is an eighty-year-old body composed of eighteen member nations. Scientists on the panel examined all data available on the population status and dynamics of the harp seal. The panel concluded that harps had likely been steadily increasing in number between the late '60s and the late '70s. Further, the panel made some recommendations to those nations which use the resource: it was suggested that in the best interests of the seals themselves, continued monitoring of the numbers of pups produced yearly, and their physical condition, be carried out internationally.

This is significant since it recognizes the fact that animals under nutritional stress produce fewer live young, and those are smaller and less likely to survive. This recommendation may have come out of fear that a ban on the commercial whitecoat hunt would result in a population explosion similar to that which took place during the two world wars. Seal health in general deteriorated and parasites flourished after a period of no hunting due to war activity in the Atlantic.

Since the ICES panel findings were made public, both Greenpeace and the International Fund for Animal Welfare spokesmen have disputed their findings. Notwithstanding these criticisms, however, the North Atlantic Fisheries Organization Scientific Council, in 1983, essentially confirmed the findings of the ICES panel.

Since then, most groups are no longer claiming that extinction is a real possibility, but have shifted focus to "moral reasons" why wildlife and wild infant creatures in particular should not be taken. In fact, claims of extinction have all diminished except by one group, whose members probably never heard of the ICES, and certainly never will through their organization's magazine.

Those on the ICES panel knew that the mood in Europe was about to cause an import ban on pelts of immature seals and were well aware of the potential which this had for impact on herd welfare. Those scientists, and McCloskey himself, probably suspected that all the sane, rational discussion in the world on this subject would not change public emotion one whit, as long as the message of seal hunt protest continued to be so widely distributed.

England and McCloskey both reported on the seal hunt with integrity, basing their impressions and opinions on the values in Newfoundland culture which they observed and on their own experience in observing seals and sealing. McCloskey and others in this emotion-charged climate of the '70s and '80s have recognized that their work is a small but steadily increasing voice which is telling yet another seal story to the public with candor and enthusiasm.

So far there has been no protest organization move to discredit McCloskey's writings. This may be because they have reached a less selective audience than does each protest newsletter, and partly because they are so disarming and credible that the task of calling them into question would not be worth the effort. Bill spent a long, hard working time out there on the ice, and he has told his side.

IV

To Kill A Seal:
Development of Agreement on Humane Technique

All of earth's conscious, alert species have in common two physical systems which must function in order that normal life can be maintained. These are the brain, centre of an intricate central nervous system, and a blood supply network. The brain controls consciousness and all unconscious but necessary functions such as breathing and heart rate, and the blood brings oxygen to it and all other structures and organs, and then cleanses them of wastes by taking various by-products off to the liver, lungs or kidneys.

If either the brain or the blood supply system is significantly damaged or destroyed, the creature will suffer a rather quick death due to neglect of its vital functions. Instinctive programming or actual knowledge of this principle has been the basis for the behaviour of all predators which have depended on the meat of other creatures in order to live. It has also been important in self-defense techniques in the natural world.

Pain is the result of any injury to a sensor of the central nervous system, and is a signal to the brain which usually results in a withdrawal of the body part which has suffered damage, away from the source of injury. A cut or burned extremity, for instance, pulls back from the offending edge or heat source, even before a conscious response has been formed. Conscious response may then be flight or aggressive retaliation in order to prevent further damage. Either response will be anticipated and perhaps countered by a predator which has inflicted damage, and which may have to fight or chase its intended prey.

Man's treatment of either wild or domesticated prey animals has been to destroy at least the blood supply system in order to effect death. In many cases, he has also elected to first destroy the brain in order to prevent damage to himself or loss of his prey. Only in the last few years of his existence has man cared about the prevention of pain in his prey animals, and come to realize that brain destruction can accomplish both this and effective death so that the body can be cleansed of blood without risk of personal injury after the head blow has been administered.

While prevention of pain in the slaughtered animal is now (for some) a secondary cultural goal, sought because it satisfies a new human need to be compassionate, it probably would not be of great importance to those societies which consume meat if it were not so easy to accomplish as a part of normal butchering. When the brain is significantly damaged by a blow to the head, the animal can no longer bite, run or fight back in a conscious manner. Loss of consciousness and pain is a by-product of this process, not the primary reason for it.

Fortunately for those pain-sensitive creatures which are our slaughter objects, it is possible to destroy the brain's pain control centre without causing the heart to stop beating, thus we can cause instant loss of all consciousness and then proceed to empty the body of the blood which would cause the meat to spoil if left therein. Death is painless and the hide and meat are left ready for use.

This is the principle used in modern slaughterhouses. It is not pleasant to watch, but one may at least be assured that the steer or hog coming down the chute to the kill floor will experience, in most cases, no more than momentary anxiety about the unfamiliar nature of its surroundings before the blow is administered. This blow used to be delivered by a knocking hammer swung by a man who stood at the head and who, understandably, became fatigued after smashing the skulls of a large number of large animals. Sometimes more than one blow would be necessary.

A more modern and undoubtedly more humane technique was the development of a captive bolt pistol: the gun would be placed to the forehead or over the ear of the animal and the trigger pulled. Instead of a bullet which might ricochet from concrete floor or walls, the projectile is a solid cylinder which comes out the gun barrel, does its damage by entering the skull and delivering a severe blow to the brain, then retracts back into the gun. It is powered by a .22 pistol mechanism and can be used all day with a minimum of fatigue or loss of accuracy in the user.

Placement of the blow by positioning of the gun directly on the surface of the animal's head is all-important to an instant, painless deathblow. The animal's body then goes through the following motions: the eyes roll back, the legs go out from under it, it falls to one side and lays inert. A few seconds may pass before the hind legs begin a violent kicking reflex caused by traumatic compression of the spinal fluid.

In the case of a domestic stock-slaughtering operation, the few seconds before the hind legs begin to kick randomly are used to place on one of them a rope or chain attached to a mechanical hoist. Then the animal is lifted vertically by the hind leg until the body swings free of the floor and its throat is slashed by the butcher who understands that both veins and arteries in that area should be severed in order that the carcass drain properly. If true, irreversible death had not been accomplished by the damagae to the brain a few second earlier, the process is now completed within twenty seconds more as all life blood leaves the body. The animal is dead.

The United States Department of Agriculture sends inspectors to each federally-licensed meat plant in the country in order to assure that all stock is slaughtered in as humane a manner as possible. The other duties of these inspectors include sanitation and safety procedures. The above procedure of death for domestic stock is rigidly adhered to. Death by such a so-called "painless" method falls under the definition of euthanasia, which is used by veterinarians in the United States and elsewhere.

Euthanasia, or painless death, may be accomplished in a number of ways, according to the criteria of the American Veterinary Medical Association. This body periodically examines methods of animal death which are in current usage for different reasons, and tests them with criteria used to decide if the method indeed results in a painless death for the animal. The results are then published in the *Journal of The American Veterinary Medical Association*. This discussion is taken from the July 1, 1978 (Vol. 173) issue of the *Journal*.

Methods of euthanizing animals with drugs and mechanical devices are examined for painlessness, human safety, and whether or not the death of the animal includes reflexes or vocalizations which may cause distress in human observers. The method in question is then judged to be acceptable or not acceptable within the above limits of observation.

Gunshot to the brain under field conditions is adjudged to be an approved method of euthanasia if an animal cannot be restrained for the blow and if accuracy can be attained. The captive bolt pistol is equally acceptable as a means of painless death. Both, it is noted in the report, result in instant death yet are aesthetically unpleasant.

The report also examines death by other physical methods, as follows:

> If a blow of sufficient force and accuracy is delivered to the skull of an animal, immediate unconsciousness results. Death should then be assured by thoracotomy, exsanguination, or other means. Because this procedure is esthetically offensive, it is not recommended where more suitable techniques are available.

Out on the seal ice other techniques have been tried and have been found less than appropriate. Gunshot death for seal pups has been found to be efficient, but highly unnecessary. Whitecoat skulls are very soft. The animals lie nearly immobile on the ice and seldom try to avoid the presence of a human. They do not seem to have a fear response, but appear quite dull-witted, even disinterested in activity around them. There is no problem with accuracy for a man with a hardwood club or Norwegian *hakapik* instrument.

There is a humane problem with the use of a gun for whitecoat slaughter, however, and that is due to their keen sense of hearing. Seals suffer if loud concussion-type noises repeatedly occur near them. Their hearing is highly developed for ease in their own underwater communication patterns. Since sound travels more easily in air than in water, repeated gunshots on the seal ice would be a source of considerable stress for any seals in the area. Since the pups are usually quite close to each other in the whelping patch, many would feel this stress before they died from a point blank shot, and such stress could be considered quite inhumane and unnecessary. There is also a safety problem with the use of gunshot on the ice. Ricochet is always a possibility.

Nevertheless, since the public has been encouraged by protest groups to object so strenuously to clubbing and bleeding out as a method of death, numerous experiments with guns have been carried out to see if seals would die as humanely with gunshot. The results were physically impressive. Brains were destroyed by one shot with no problem of recovery to a conscious state in which pain could be felt. There is no doubt that guns work well, except for the significant noise factor. However, it is illegal for a Canadian citizen to have or carry a handgun or ammunition for same. If seals were to be killed with pistols these would have to be issued to individuals for the season, along with counted ammunition, and all returned to the government at the end of the season with an elaborate accounting.

It is difficult enough to move across rough ice and pull a heavy tow of seal pelts and meat without worrying about safe gun handling. A handgun would pose a hazard to a man every time he fell down, and if he should fall in the water it would probably freeze up and jam. A long gun would be no better, and probably worse. To safely carry around a rifle without damaging it while falling, jumping or slipping would be a tiring task. Each man must still be able to carry out other sealing functions, such as pelting and hauling, and the extra weight and nuisance of a long gun would be considerable, especially when the man carrying it would not be able to rely upon it to pull him out of the water (a further function of his *hakapik*).

A sealer knows his club will be safer than any gun could ever be, and will never give him a lot of trouble over paperwork which any handgun surely would. In addition, the Seal Protection Regulations state that every shot seal must still be clubbed before it is skinned in order to assure braindeath. The advantage of a gun is thus seen to be nil. Acoustic stress and human safety present considerable problems. Silencers would perhaps be an expensive solution to the first, but would not improve human safety which is more important.

While it might be possible to inject a seal with a drug which would kill it humanely, such an animal could not then be used for human consumption. Injection would have to be intravenous, presumably through a flipper, and two people would have to restrain a struggling, bitting, up to eighty pound animal in order to find a vein not covered by inches of blubber, and then hold it still long enough for the drug to enter the heart and stop it.

Heart stoppage would prevent the bleeding out which is necessary to ensure high pelt quality. The result would be a highly stressed, frightened animal which would not die for several to many minutes from the time of first restraint. Its pelt would be highly susceptible to "ice burn" because of retained blood and body heat. Ice burned pelts are worthless. The meat would be unfit due to the drug. The men administering it would be angry, bitten and exhausted. All drug supplies would have to be accounted for. The method would be totally unacceptable and very expensive.

The following is taken verbatim from the 1981 issue of *The Sealer's Guide*, a handbook for licensed sealers issued, with licenses and other instructive materials, by the Canadian Department of Fisheries and Oceans:

Use of the Club and Hakapik

a. Only the regulatory club and hakapik can be used to strike a live seal.
b. Seals must be struck only on the forehead.
c. The regulations specify that, in order to stun a seal, a sealer must strike it three times on the forehead with a club or hakapik, or until the skull is crushed.
d. The sealer must then check to make sure the animal is unconscious. This is done simply by touching the eye; if it blinks the seal is still conscious and must immediately be struck on the head to render it irreversibly unconscious.
e. Seals that have been shot must also be struck on the forehead with a hakapik or club. Before an attempt is made to skin the seal, the sealer must make sure that the seal has no blinking reflex when one of its eyes is touched and that it is in an irreversible state of unconsciousness or that its skull is crushed.
f. When the animal is unconscious, the seal must be bled immediately by cutting the major blood vessels to the fore-flippers.

Skinning

a. A seal must not be skinned until the sealer has checked to see that it is in a state of irreversible unconsciousness. It must be glassy-eyed, have a staring appearance, have no blinking reflex when the eye is touched and its muscles must be relaxed.
b. The sealing group leader must see to it that the members of his team immediately bleed out the seals they have stunned, and skin them without delay; at no time should there be on the ice more than ten dead seals which have not been skinned. Strict observance of this procedure is essential to avoid iceburned pelts which are worthless to anyone.

The *Sealer's Guide* was prepared in order that the Seal Protection Regulations, a piece of legislation covering all sealing in Canada, be most easily understood and adhered to. The *Guide*, and a movie made in both French and English, serve as mutually reinforcing instructional aides in classes which are given prospective sealers before licenses are issued.

Ordinarily there is more or less constant supervision of all sealing done by both ship-based and land-based sealers. This is accomplished by the presence of trained Fisheries officers on the large ships and by the use of helicopters, from which officers can spot small boats and longliners and drop in to observe whether or not regulations are satisfactorily met. A special contingent of officers is assigned to duty in the Gulf of St. Lawrence, on the Magdalen Islands, during seal time each spring. These men live in a motel headquarters there for the duration, along with rotating flight crews.

This writer observed great interest on the part of such officers to inspect each and every sealing team on the ice, and they were constantly instructing their pilots to circle, hover, and usually land near the boat, in order to check for current licenses and watch the men at work. The sealers seemed to expect this sort of behaviour, and in return for the visit also expected any news of near-by seal patches which helicopter passengers could see more easily than could anyone down at ice level.

Lengthy discussions in French were carried out between sealers and officers on everything from why I wished to take pictures to the fortunate large size of the few seals which they had managed to find. Ice conditions were extremely rough and the work was hard. Everyone had on his sealing badge and licenses were apparently in order. Seals in that area were being killed with clubs and routinely bled and pelted. There did not seem to be any difficulty in achieving the quick death required by the Seal Protection Regulations.

Sealers living on the Magdalen Islands are experts in their trade and have a long tradition of dependence on the resource. There are always some seals to be had, and usually the ice brings them within reach. Once in a great while, however, nature changes the game. The wind and ice conditions may be such that no boats can force their way through to the seals. Only a few, whelped around the edges of the ice, can be obtained.

Rarely, the ice is blown far to the south so that seals are scarce in the Magdalens but common on the north shores of Prince Edward Island. In 1969 and 1981 this happened, along with the formation of ice which was very weak and spotty, hardly supporting the seals which happened to be in that area. Small thin pans of ice, thinly scattered across the sea, caused the unthinkable to happen: some harp seals actually came onto land to whelp rather than give birth on such poor ice. Farther out from Prince Edward Island the ice was better and those seals which used it had better luck. But such information cannot be shared by animals such as these, and a large number were in trouble.

In 1981 the people of Prince Edward Island, however, considered it a rare and fortunate happening. Those who were fishermen wanted to be issued licenses to go out and take seals, even though they had not had such an opportunity for the past dozen years. A windfall resource had suddenly come in; there was money to be made when none had been expected and there was a great deal of genuine excitement.

Fisheries Department personnel made the decision to give prelicensing instruction to some two hundred and forty sealers who insisted on the opportunity. The movie A-1, which includes instruction on humane killing and quality pelt production, was shown and there was discussion of proper killing and skinning technique. Unfortunately, however, most of those men who obtained new licenses were entirely inexperienced due to lack of previous recent opportunity. This inexperience was subsequently witnessed by a large number of people. The resultant publicity flap was heard around the world, and as one Fisheries spokesman later expressed it, twelve years of effort at establishing the credibility of Canadian sealing went down the drain.

On Sunday, March 8, 1981, the seal ice blew right up to the shore on Prince Edward Island and people standing on shore were able to watch men trying to kill the pups by clubbing them. The crowd contained newspersons and a few members of the Animal Protection Institute from California. According to the report of Colin Platt, Regional Director of the World Society for the Protection of Animals, the landsmen sealers were largely unable to carry out sealing in the approved way and were seen to be delivering repeated club blows to pups, then flinging their reflexing bodies into the boats without first bleeding them out. Platt noted that no Fisheries officers were out on the ice pans doing anything about this violation but that some were observing from the air. Since the ice was so poor it was difficult for a helicopter to safely land.

On Monday, March 9, Colin Platt himself, at the request of some sealers whom he observed working on the ice, demonstrated humane clubbing, bleeding out and skinning techniques, and then showed them how to properly stack their pelts to prevent ice-burn. He was later commended highly by Fisheries Department personnel who recognized that he had acted in the best interest of seals and men and had done his best to correct a very bad situation.

Before their participation in the hunt was called off, some two thousand seal pups were taken in two days by landsmen who had virtually no experience in sealing. By Tuesday, March 10, the landsmen's hunt had been cancelled. Sealing was still allowed offshore by crews of the *Brendal* and the *Techno Venture*, men who were old hands at clubbing and bleeding seals.

Numerous violations of the Seal Protection Regulations had been reported by foreign humane observers, members of the Committee on Seals and Sealing, and by Fisheries officers. The hunt by landsmen was stopped for this reason. The fact that the public had been able to observe sealing at its worst was incidental to the shutdown, but of course protest groups and press reports credited public outrage for the closing of the season for landsmen.

There was, unfortunately, illegal seizing of camera equipment and film of the hunt fiasco by an officer of the RCMP. The girl from whom the film was taken suffered a bruised hand in a scuffle, the details of which probably never will be known from an unbiased viewpoint. She belonged to the Animal Protection Institute and that group published her version of the incident as an outrageous, unprovoked attack on her by a male police officer who wished to suppress her right to film in a public park. The film was destroyed so her pictures of men ineffectively beating seals were lost.

In retrospect, the PEI massacre (as it became known) is the only really black mark on the humane record of Canadian sealing. A mistake in judgement was made in the licensing of people who were inexperienced and who had too few experienced leaders among them. Mistakes were made about the normal civil rights of the public as they stood on shore, not on the ice, in close proximity to seals. It was a time of extreme tension and fear because there was no precedent for proper action which could be followed. There is no doubt in most people's minds that mutual lack of respect was a factor in police-crowd relations on those two days.

Those few violations of the Seal Protection Regulations which were observed on the *Brendal* on March 8 were reported and the men corrected. Since regulations were first enforced in 1965, over thirty sealers have been fined, lost their licenses, and removed from the ice for the remainder of that season. Yet claims have regularly been made by some protest organizations that no arrests for cruel treatment of seals have ever been made and that the regulations were a farce designed simply to control protest activity. It would appear that this is not true. A list of arrests for specific violations of the Regulations was made available to this writer as soon as it was requested, with no stalling or apologies. A Fisheries Department report of the PEI fiasco was also made available, along with Colin Platt's candid account of his two days observing sealing in 1981. Fisheries people freely admitted to being human and to having made some mistakes.

The landsmen on Prince Edward Island probably won't have an opportunity to obtain sealing licenses again, unless they should go to Newfoundland and apprentice under experienced men. This is entirely unlikely since it would be expensive to travel and would require social relationships which do not already exist. The chance of the seals coming to Prince Edward Island again on a regular basis is slim and unpredictable. The only sealing which may take place off PEI will be that done by roving ships manned by those who have been sealing for a lifetime.

The 1981 affair was blown out of all proportion, in the words of Fisheries, by the press, and by protest organization advertisements and rhetoric. The result was a widespread public belief that sealing is always inhumane, inefficient, and ugly. Those fishermen who regularly go sealing and are proud of their tradition suffered from knowing that this part of their livelihood was jeopardized by bad press and the embarrassment of government officials.

During the whole affair the public never heard about the continuing gun trials which were compared with death by clubbing by men on the *Brendal*. There was no press release on the efficiency of clubbing in those comparison trials which was disclosed by autopsy of those seals tested. It would not have been timely anyway.

The enforcement of regulations which ensure that sealing is a humane activity is an expensive procedure. However, this enforcement is apparently carried out with vigor by able Fisheries officers who go to great lengths to enforce humane killing methods among licensed sealers. Officers with whom I spoke were aware that they were under foreign scrutiny, but according to sealers themselves, the diligence of the law is a constant factor in the spring. Helicopters are leased and pilots are paid by the companies which own the machines. The pilots have to be experienced in "flying seals" so the officers can carry out their enforcement duties and biologists can land on the ice to tag and examine seal pups of two species. There is no room for mistakes when flying over sea ice, miles from land.

Research goes on from dawn to dark as soon as the seals come in to whelp and those animals which are tagged are not supposed to be taken that year by the hunters. Those who return a tag from an animal of some previous year receive a small monetary reward for doing so. The scenario is not unique among wildlife management programmes in the western hemisphere. It has been compared with the management of many other species, both land and marine, and world opinion is that Canadians have done an admirable job of research and management of harp and hooded seals.

This writer sought the opinion of the National Wildlife Federation as a part of her research on the reputation Canada has for humane control of the seal hunt. The organization was also asked for its opinion on Canada's record of herd management since so many protest groups have questioned both humane killing technique and seal population status. The reply to these questions is as follows:

> National Wildlife Federation does not oppose consumptive use of wildlife provided harvests are carried out under scientific management programs that safely guard animal population and that the most humane methods possible of killing are employed. Based on available data, both of these factors are being met by Canadian regulations of the harp seal harvest. We appreciate your objectivity in researching this issue, and look forward to seeing your final product.

The International Association of Fish and Wildlife Agencies was similarly querried for its position on the harp seal harvest. Their reply of June 9, 1982 states:

> We have confidence in the data and management competence of our Canadian neighbors and in their judgement as to whether the harvest level may be damaging to the Harp seal population. Also, the method of killing has been examined by a number of independent groups and found to be the most effective as well as the most humane.

This organization does not maintain a file of all data collected on any species, but its members are aware of the great volume of research which has been carried out on the harp seal in Canada since 1950. While the Internatonal Association of Fish and Wildlife Agencies has not taken an official position on the harp seal harvest (this is not one of their functions), they are an internationally-recognized body of scientists who make countless decisions regarding the funding of state and private wildlife research projects. Thus they are in a position to know the status and stature of wildlife research and to comment when asked.

The Canadian Federation of Humane Societies has said that "As far as we are concerned, the present regulations ensure that the best possible methods of humane killing are adhered to.... Humane killing is not an issue."

Dr. Harry Rowsell, D.V.M., Ph.D., Department of Pathology, University of Ottawa, and Canadian Council on Animal Care, has personally examined the carcasses of seals which were slaughtered in the normal way by men using clubs. His impression was entirely favourable: "I have examined the craniums of thousands of seal pups and I have never observed one that did not have massive hemorrhage in the brain, which is an indication that the animal was rendered unconscious and therefore incapable of feeling any pain."

The slaughter of seal pups looks very much like the slaughter of any domestic stock which are killed by the same basic method. It is unpleasant for an observer to watch. If that observer does not understand that the reflexes and twitches are made while the animal is unconscious or actually braindead, then naturally such a person is apt to protest that the killing is unbelievably cruel. All those observers who have been competent to judge the method have agreed that it is indisputably humane. The victims of cruelty have been those who are poorly informed.

The matter of physical pain to the slaughtered animal has been investigated and discussed endlessly in the last twenty years. The subject of physchological pain and trauma to the adult seals which on occasion see pups being killed has not been so tidily answered. This should be considered here since to ignore this within a discussion on cruelty to seals would be inappropriate.

Many observers of the seal hunt have remarked that some female seals do seem to be aggressive in defense of their own pups, while others merely slip into the water and disappear. Later, when the killing is done and the men are gone, some seals come back ont the ice and can be seen sniffing the carcasses. Much has been made of this behaviour by some writers who have claimed that the dams weep and appear "heartbroken" over the loss of their young.

One must understand these animals in order to be able to assess fairly whether or not they feel the level of loss which a bitch dog often exhibits if her newborn pups die or are taken from her. A dog will cry, whine, and mope about if her nursing young are removed. She may be very aggressive in defense of her young from the beginning, and it is clear to anyone that a dog, at least, can feel emotional pain. Seals are not demonstrably dog-like in their normal maternal behaviour, nor is it agreed that they appear to have the amount or kind of intelligence which has evolved in any canine species.

A dog normally nurses its young for a minimum of three weeks, and longer than that in the normal state, while pups mature to the point where they can eat regurgitated food, and later, hunt. Maternal care involves much more than nursing as life training is involved in a bitch-pup relationship. This is not the case with harp seals which nurse their young for ten days then refuse to suckle the pup after that time, although the dams may occasionally lounge on the ice nearby, resting before breeding again.

The open season on seal pups starts when authorities have been informed that a majority of dams have whelped and that most pups have developed white coats in place of the yellow lanugo or birth coat, which lasts about three days. By the time commercial sealing starts, many pups are at least ten days old, although there are always some older and some younger on the ice when the men arrive.

A newborn "cat" seal is not worth much and would be passed by if seals were plentiful in that area. Its mother would be in that category of dams which spend the most time on the ice with the pup, nursing several to many times a day. The large and fat ten-day-old seals are those which are newly abandoned. Most of their mothers are at that part of their cycle which dictates that they stay in the water seeking mates. In a few instances, during which this writer saw a patch of over twenty thousand whitecoats, there was a harp dam which hung fast by her pup and repeatedly returned to it even though the research team was close at hand.

Those harp seals which gave birth for the first time are more apt to stay with the pup longer than is necessary for its nourishment. Number of births a seal had undergone cannot be determined with great accuracy, but if she has had only one pup this can usually be determined by post-mortem examination of such a seal's ovaries and by the fact that her coat colouration is not that of an "old" or mature harp. The seals which I saw vigorously defending their young were rare in a crowd of twenty thousand animals; only three were noted during two days of "flying seals" in 1982 when the hunt was still going on.

Bill McCloskey, another American writer who has spent productive time on the ice and who has observed the hunt more than any protest group members, saw men pass by seal pups which were still attended by their mothers. It recently has been made illegal to kill an adult seal on whelping or breeding ice, and to remove a small pup from a defensive mother would have been more bother and danger than it would be worth. A defensive harp dam can travel across the ice more rapidly than most people with cleats on their boots, and a bite from that cat-fanged mouth would be no laughing matter. A harp dam may weigh as much as three hundred pounds and is a powerful and nasty appearing animal.

Some harp dams will stay by their young while men are near and it has been suggested in the *Sealer's Guide* that such attended pups not be taken. The dam may not be killed, even if she is defensive. This latter regulation, along with the suggestion that the pup be left on the ice while she is there, should help to alleviate the stress or maternal upset which used to happen occasionally durng the hunt.

Of course, those who have claimed that seal dams weep because they mourn the loss of their young have been either ignorant of seal physiology or guilty of deliberate and inexcusable untruth. The latter often appears to have been the case. As noted previously, all seals produce copious thick tears which are secreted all the time and which appear to gush forth every time one exits the water and is exposed to air. The misrepresentation of this physiological response to a change in environmental surroundings, and the use of it as a gimmick with which to win sympathy for a "Save the Seals" fund-raising campaign, has been noted with disgust by scientists for two decades.

One who stood for hours on the ice and watched female harp seals return to sniff the carcasses of slain, skinned pups would have to feel some wonder at whether the animals were upset or merely interested in the presence of this new and different object. The animals themselves do not appear to spend much time at this behaviour, or to hang around a particular carcass unless it lies in the spot where the dam was accustomed to coming out and resting anyway. It must be remembered that these animals had developed a territorial feeling for the individual whelping patch and for a dam to return to her own and stay there to rest may well be an expected behaviour, not a sign of grief.

In conclusion on this point it is reasonable to expect that an animal which loses its young on the eighth day of its existence, or on the tenth day, would return to the spot more frequently than she would if she had not had a territory of her own to begin with. However, since she would have abandoned maternal care of such a pup by the tenth day anyway, how strong can her feeling of loss be? There is no evidence that grief or loss as we know it exists for the harp seal, regardless of highly-coloured claims to the contrary.

The wider question of basic cruelty to the species as a whole through hunt activity which concentrates on the very young is a further focus of some protest statements. Greenpeace Canada has taken the position that it is immoral to hunt "nursing young" and to interrupt the tranquility of the maternal haven with clubs and knives. Brigitte Bardot has used this same theme in her protest of the seal hunt, as has author Richard Adams (Watership Down) during appearances in *The Rites of Spring* and *The Hunt Without Pity*. Bardot's point seems to be that it is somehow "unfair" to "hunt babies," with the implication that they are defenseless, unlike, presumably, large male prey. Richard Adams spoke about the inhumanity of taking "cubs and pups who are still at their mothers' teats."

Of course, the filming of these two productions took place before

1980, at which time it became illegal to kill a defensive harp dam or any other adult animal on the ice. In fairness to both the above celebrity protesters, they may have seen an IFAW- or Greenpeace-produced film which depicted the taking of a pup and the dragging of it across the ice while a helpless female seal followed and then, according to the narrator, was herself killed. It is not known whether or not the scene was genuine or staged for the camera, but one may assume that such a scene or the description of it probably influenced these two people to make their remarks.

Incidentally, the regulation which prohibits the taking of adult seals is a conservation measure, not one which may be seen as a concession to protest statements. It has long been recognized that adult seals have a lower mortality rate and a much greater statistical chance of influencing herd growth than do animals under one year of age. This is not unique to seals, but is common in species which are long-lived and which take a rather long time to reach sexual maturity. Breeding animals must be protected in order that herd growth be maintained or increased. A large proportion of young will die of natural causes or can be taken by human predation without causing herd depletion.

This is a principle which has been expressed by the World Wildlife Fund policy statement on harp seals issued in 1982. The statement calls for continued monitoring of harp and hooded seals by Canada and other nations which have access to them. It states that WWF feels a moral responsibility to the herd as a whole, not just to individual members of that herd, while recognizing that humane killing has to be employed in the taking of the animals. It notes with concern that the harp population seems to be growing and that the age structure of this must continue to be monitored. It notes that "to ensure a healthy reproductive capacity in the herd, no more than 20% of the animals taken should be over one year of age. WWF wants to see sufficient regulation of the hunt this year to make sure that too many older seals are *not* taken." The organization does not want its position to be interpreted as "being one that promotes or is unconditionally for the hunt. That is why we have tried to be specific about what our conditions and concerns are. Our position is, by definition, closely tied to what we judge to be the best population data available on the biological status of the herd...."

As long as there has been a hunt for harp seals on the eastern coast of Canada there has been concentration of effort on taking whitecoats, especially in the years since the Second World War. Seal pups have traditionally been taken with club and knife and adults have traditionally been shot, in recent history. In spite of well-financed international protest about cruelty inherent in the methods, there is universal agreement among veterinary scientists and biologists that clubbing with bleeding out is the most humane method ever proposed or tested. Gunshot of pups is a close second, but the noise factor and the human safety factor are significant drawbacks.

It has been suggested that basically it is not the *manner* of taking, but the *fact* of taking infant animals which has become a focus of protest. In light of all the internationally recognized valid data on harp seal population structure and manner of population explosion during times of non-predation by man, it would seem that the general public needs to be thoroughly educated in this matter in order that protest and loss of legal markets not adversely affect the well-being of harp seals, as a species and as individuals.

In conclusion, while many people feel a deep love for wildlife, that love alone is not an adequate basis for responsible decision-making about such things as hunting in general, selective hunting of certain age groups, or the closing of legal seasons. Many feel it is unfortunate that the animal rights/environmentalist protest movement, a growth industry in the last two decades, has spent most of its time adversely influencing public opinion on the taking of wildlife in general and on the harp seal issue in particular.

It has been argued that everyone has a right to his or her own opinion about such matters, yet an uninformed opinion which results in harm to a species and to thousands of humans as well is not defensible. The issue of cruelty to harp seals has been one nurtured by a combination of ignorance and clever fund-raising. The resulting public-generated legislation in Europe and the United States is destined to have an effect on the biology of many species for some years to come.

V

Saving the Seals:
History of Seal Management in Canada

Whenever a wildlife species is considered a resource for humans it becomes an object of governmental concern. Its abundance or scarcity is monitored by professional scientists, either in the direct employ of the government, on contract with it through a university centre, or both. In addition, the government may also cooperate fully with outside, independent researchers in universities by allowing them full access to the resource and encouraging their study through grants or other financial aid in exchange for the data which they gather.

Professional wildlife biologists are fulltime specialists with advanced degrees in their fields. While their concerns may cover whole ecosystems, their specific employment usually depends on problem-solving in certain areas. Species are counted, and the impact of each is measured as part of the local ecosystem. Interspecies relationships are recorded and examined as conditions change through time. The object of all this study is to gain some control over populations within a particular area, and to achieve some predictability about how compatible these can be, now and in the future, not only with each other, but with the humans who also inhabit the territory.

Government sponsored wildlife management programmes are traditionally concerned with the welfare of all species within their jurisdictions. Creatures being managed must be protected from even the threat of extinction, and at the same time prevented from becoming a nuisance or a scourge to one another and to the people in the area. A balance of all living creatures is the ultimate goal.

Government wildlife management policies usually reflect the wishes of those citizens who are closest to the resource being managed. In many cases, these people are either hunters and fishermen, or farmers and ranchers, all of whom are affected by the life cycles of land and marine wildlife. Their interests have come first on any priority list which is kept for governmental attention because these citizens are those who have the ears of their representatives.

Thus, accountability to the consumers of wildlife, and to those whose livelihood is adversely affected by wildlife, has been common in recent history. Ranchers want advice and help with control of wolves, coyotes, mountain lions, prairie dogs, and eagles in their environment. Fishermen want abundance for their nets, and seals for fresh meat and extra income. Native peoples need an abundance of seals at all times of the year for direct consumption and because the sealskin which they bring in to the store is their only source of cash. There is nothing else which can be exchanged for food, warm clothing, medicine, fuel for boats and snowmobiles, and ammunition.

The traditional form of wildlife management which has developed over the last fifty years has come about in order to satisfy man's needs, but the end result has been in many cases to keep many wild species afloat, neither in danger nor in overabundance. Commercial hunting of all species has been either outlawed or modified, with strict selective harvesting rules meant to protect the creature as a continuing resource.

There is evidence that Canada has always had the best interests of her people and her seals at heart in thirty-odd years of attempting to understand and manage them. While some factors, such as offshore commercial fishing operations by other nations, have been out of her control, she has at least had a voice in the management of those resources through her participation in NAFO, the North Atlantic Fisheries Organization, which conducts research on fish biomass and which manages the fishery through quota setting for all participating countries.

The seals themselves are a reflection of the management policies which seek to influence their numbers; when herds are too dense the condition of individual animals is affected. A simple measurement of this is blubber thickness; in times of food scarcity it is less than when the fishing is good. Other indicators are relative parasite load, fighting wounds on the males, and the age at which females attain sexual maturity. Pup mortality, incidence of stillbirths, and pup weight at birth are all routinely measured by biologists.

If seal condition is seen to deteriorate, indicating over-population for available energy supplies in the ecosystem, the answer may be either an increase in seal hunt quotas, or a corresponding decrease in the amount of fish which humans are allowed to take. Neither is an easily accepted answer. Each solution is met with opposition from non-scientific interests.

By the 1980s western Atlantic harp seals were internationally recognized in scientific circles as healthy and thriving. The world public, however, had been led to believe the opposite; that the seals were endangered through a lack of governmental concern, and that continued hunting would inevitably cause their demise as a species. The following is a brief history of seal management as it has developed, and the reasons why it has been such a success, for the seals and for those who have depended on them.

There was no seal management in Canada on a federal level until after Newfoundland became a province in 1949. In 1950 extensive government-sponsored research on harp and hooded seals began, as it was common knowledge that the "seal fishery" was very important to the entire east coast, and especially to the people of Newfoundland.

It was also well known that the seal population had fluctuated rather widely during the history of its exploitation. Since there was general agreement that the resource was a vital part of life in the Maritimes, where literally everyone's livelihood depended, at different times of the year, on either sealing or fishing, the decision was made to ensure that seals continued in abundance.

In 1950 the body of knowledge about harp seals included the fact that females all came onto the ice to have their pups within about a two-week period at the end of February, nursed for only ten days, then abandoned their young to engage in the annual breeding activity.

It was decided that in order to "manage" these seals, their numbers would have to be known with a high degree of accuracy. Therefore, attempts should be made to count seals when they were on the ice. The first aerial photographs of seals on the whelping ice were taken in 1950 and the problems with using photos to estimate total numbers began to be recognized.

The photographs always had to be taken from the same altitude so that, superimposed on a grid, they would always be covering an area of uniform size. Since seals are never distributed evenly over the ice, the entire area had to be photographed and the plane had to fly in a strip pattern over hundreds of square miles. Terrible weather conditions, including fog and storms, made uniform coverage nearly impossible.

Other problems included the fact that not all coloured adult seals on the ice during the whelping season are fresh dams. Some males and some juveniles from past years would be present; therefore, extrapolating numbers of pups from numbers of coloured adults would have to be adjusted from the sex ratio of adults observed from ice level. The time of day and cloud cover also affected the length of time females would be on the ice with their young. Obviously, seal management was not going to be easily successful, nor very economical, until such problems were solved with "ground truthing" checks by personnel on the ice.

Also, before the use of ultraviolet (heat sensitive) film, white pups could not be recorded as they did not show up against the background.

Between 1950 and 1970, however, international cooperation among Canada, Denmark, Norway and the Soviet Union resulted in refinement of aerial survey, ground truthing, and statistical model techniques. All these nations contributed to this research since all were engaged in sealing and had a vested interest in the continuation of the resource. Data on age of animals at the time of harvest, general physical condition, sex, state of gestation, weight, parasite load, evidence of disease or injury, all were pooled into an international collection of knowledge. In over twenty years of research some 3,600 published papers were distributed among scientists internationally and computerized files were established. Seal research had become a complex field.

Naturally, not all scientists were in agreement that methods used in population estimates were reliable in the same degree and in the 1970s considerable discussion over whether the population was declining or growing resulted. The difficulty was in estimating natural mortality for the herd in general.

The first quotas were set in 1971, at which time the allowable catch was 245,000. Due to disagreements over herd strength, however, this was reduced to 150,000 between 1972 and 1975. In 1976, the argument continuing, Canadian authorities decided to take an extremely conservative approach to management and lowered the total allowable catch to 127,000. Since then, however, scientists have come to agree on an overall herd mortality rate of eleven percent, much lower than some had earlier suggested. Accordingly the quotas were raised thereafter, it having been decided that this would not decrease herd strength and would actually allow for a gradual increase.

In order to "manage" a population of any creature, it is necessary to know age and sex ratios within the species, its rates of birth and death, and those factors which affect growth and decline. Although this knowledge has come to maturity fairly quickly in the case of the harp seal, it is knowledge that has traditionally been restricted to an elite group, those scientists who participated in the process, and those government administrators who arranged the funding for the entire research programme.

The ordinary person, for various reasons, is usually unaware that wildlife is constantly being managed by professional biologists. There is no emphasis placed on this in the public school system, nor is it a part of the folklore of modern western civilization. Techniques and goals of wildlife management are sometimes known to hunters and farmers who feel a personal concern, or to insurance companies which have to pay the damage claims of property owners. Wildlife species often cause car or land or crop damage. But the urban common man or housewife or career-oriented person seldom gives any thought to wild animal or bird abundance or scarcity, since these concerns are not a part of daily life in modern society.

There are only two types of situations which have, historically, caused the public to demand a voice in the management of wildlife. The first is that in which the wild populations are impinging on space and other resources claimed by humans. The second is a perceived threat to any species which is valued for some reason, not necessarily economic. Both of these situations have been recognized as legitimate factors in harp seal management ever since sealing was first an important part of human livelihood off eastern Canada.

Even as early as 1895 laws were passed in Newfoundland which prohibited ships from making a second trip out to the ice. This measure protected animals of breeding age. The hunt traditionally concentrated on the whitecoated pups since they were on the ice and easy to approach and kill. These pups were taken for their fat and for their skins which were used more often for leather than for the fur trade. Second trips would have concentrated on the adults since the pups would have been weaned and dispersed by that time.

Before 1945 the oil from these animals was more sought after than their skins. After 1945 petroleum products became more easily available and seal fat and oil were less in demand. Seal pelts became prized for their fur, and trading of them for their valuable fine-grained leather continued.

Seals were not hunted much at all during the war years and the herds grew explosively during that time when human predation was not a factor. After the end of the Second World War, however, ships headed for the harvest again each spring and, except for the restriction on second trips, the take of seals was again largely unregulated. The market encouraged a heavy harvest, for both fur and leather. There was a continued outlet for seal oil in products such as machine lubricants and cosmetics, and this combination of demand promised to keep the annual spring seal hunt going indefinitely.

After 1964 several events took place which were to have an impact on the scientific management of these animals, and on wildlife management in general thereafter. The first was the production of a film about the hunt and about the government-sponsored research on seal population growth and conditions.

Although this film alleged to tell the public the truth about sealing, it was later disclosed that scenes showing killing methods used in the hunt had been staged and were not common practice. This film, produced by the Artek Film Company and purchased by CBC for general television viewing was *Les Phoques de la Banquise*.

The Artek film was shown widely to the public in Canada, the United States and Europe. Some scenes of seal slaughter were explicit, and some of those which were staged showed a man cutting a live adult seal with a knife, without having shot or stunned her first. This man, who had been hired by the film makers to commit this act for the camera, was later requested to testify under oath in the House of Commons about the entire incident. The hearings were held by the Standing Committee on Fisheries and Forests in 1969, four years after the film had been produced.

It was disclosed that this man, a native Magdalen Islander who had been acting as a guide for the film crew, had been hired to skin a live seal for the camera without first causing it to become unconscious in order to produce a dramatic and unforgettable scene. Understandably, the result was immediate outrage on the part of the public which witnessed this act.

Les Phoques is a historic document because it was the first film to show the clubbing, bleeding out and skinning of small whitecoated pups. Although no actual cruelty to these seals was depicted, the effect was nevertheless horrifying to a public which did not begin to understand the effectiveness of the method being used.

Although the humane killing of seals will be covered in a later chapter, a brief description of the process will be given here in order to illustrate that the killing scenes in *Les Phoques* were deliberately misleading.

Small whitecoated pups are seen lying on the ice, blinking and slowly floundering about, as is their usual habit. A man with a club approaches and a close-up shot of the seal being struck on the top of the head is seen. The seal is struck a few times, rolled over onto its back, and immediately cut with a knife between the front flippers. It is then slit from chin to tail, and blood gushes out its underside. It thrashes somewhat, as though still alive. This scene is typical of normal sealing, and while upsetting to anyone who does not understand humane butchering methods, it is not a depiction of cruelty.

Later scenes, however, do not show typical sealing and were staged for effect: a group of white pups, which have been struck and bled, are seen to be on their bellies, not turned over on their backsides to bleed out. These pups are filmed in a pool of blood on the ice, and pup voices are dubbed into this scene. The pups' eyes are open, as are the eyes of all seals or other mammals killed by any trauma.

This scene, of pups which appear to be alive, is perhaps the most dramatic and upsetting of all. It is not one which illustrates normal sealing practice. The uninformed person who sees this notes that the eyes are open, the pups are in the "alive" position, pup voices are heard, and the animals appear to be still writhing in their own blood. The illusion of animals crying in pain as they bleed to death, after having suffered apparently ineffective club blows and fatal knife wounds, is shocking and offensive in the extreme. It is probably this scene in *Les Phoques de la Banquise* which was the stimulus for all subsequent public protest about the seal hunt in general, and the manner of taking these animals in particular.

This information is necessary in a discussion of the history of seal management because the participation of a poorly-informed public became a factor in future management decisions as a direct result of the impact of this film, even after the film itself was publically revealed to be fraudulent.

The film was shown from 1965 through 1969, March through May, before the hearings were held, and up to the time of the hearings there was no public disclosure of the deliberately fraudulent nature of the production. Everyone who saw it believed that cruelty was a common part of the seal hunt and something that should be eliminated.

The narration of the film also suggested that harp and hooded seals would suffer serious depletion, to the point of becoming endangered or extinct, should the hunt continue to take animals in the hundreds of thousands, without scientifically established limits.

Although the government of Canada had been conducting responsible research on seal population trends and sealing methods, the general public was entirely unaware of the significance of this to seal management, and indeed, felt that no responsible basis for the continued hunting of seals had been established.

The first noticeable public participation in the management of harp and hooded seals began as soon as the Artek film was widely shown. Humane concern about the methods used in normal sealing resulted in investigations into those methods by the Canadian Audubon Society and by the Ontario Humane Society.

Both organizations sent representatives to the hunt, and these people all witnessed the killing of seals under normal hunt conditions. They understood that animals which are struck on the head often undergo intense yet unconscious reflexive action. They saw this effect in seals which had been clubbed, and noted in their reports that the animals appeared to be entirely unconscious before being bled out to complete the death process.

Thus, the first humane societies which undertook the responsibility of looking into the seal hunt did so in response to public interest in the conditions of slaughter, and were able to report that seals were not dying in pain, were not being skinned alive, and were not, by any recognized standards of humane treatment, being cruelly tortured.

Nevertheless, the Artek film continued to have an international effect on the public perception of the seal hunt. Canada was seen as a nation which condoned great cruelty, and which was contributing to the demise of yet another species of marine mammal. An echo of this sentiment was heard from another "eyewitness" to the seal hunt. Brian Davies himself had gone out to see seals with permission of the Canadian federal government. Davies' version was to coincide with that of the Artek film, and he has never wavered from that version in public description of the hunt, and in public denouncement of it.

According to Brian Davies, seal pups are commonly skinned alive, conscious of horrible pain and torture until their lives are snuffed out by blood loss, and by the shock of losing their insulative skin. Davies began a "Save The Seals" campaign which was first administered through the New Brunswick SPCA. Later, he formed the International Fund For Animal Welfare, and began his own worldwide campaign to "Save The Seals."

By 1969 Davies was preaching the evils of the seal hunt to a public which was more than willing to believe him. He took members of the international press, including television journalists, out to see for themselves the "torture" inflicted upon "baby" seals. Their reports, filmed and narrated for television news broadcasts, reached millions of people internationally. The public believed what it saw and heard, and sent millions of dollars to the IFAW in support of its campaign.

Davies then convinced animal lovers that they could have a significant impact on seal management policy through a massive letter-writing demonstration of their concern. The offices of provincial fisheries ministers, as well as the office of the federal minister of the Department of Fisheries and Oceans, began to receive thousands of cards and letters in protest of the seal hunt. The public wanted it stopped.

In 1969 all of this furor about sealing caused the third stimulus to public protest to come about: a book about the hunt which had first been published in 1924 was discovered and republished, with a new and very biased editorial introduction.

George Allan England's *Vikings of the Ice Floes* was now reprinted as *The Greatest Hunt in the World*, complete with photographs. Nothing could have been more timely than this explicitly descriptive work by an American who had been there himself. The editor of this new edition was astute enough to take advantage of the current mood of the public towards sealing. His introduction reflected that mood and reinforced it. In retrospect, there can be little doubt that the Artek film was responsible for the tone of this editorial introduction. Again, the result was enhanced credibility for the seal hunt protest.

Les Phoques de la Banquise must be credited with having had a lasting social impact. Even though its main scenes were staged, and much was not only incorrect but deliberately misleading, the message which it conveyed has lasted and has spawned more unjustified protest than anyone could have imagined.

Brian Davies' successful "Save The Seals" campaign began as his own strong reaction to the Artek film. The message touched him, and public attention to *Les Phoques* convinced him that people would always be moved by the sight of dying seals. By the early 1970s those who had seen the film, or had heard Davies' version of the hunt, could also read *the Greatest Hunt in the World* and believe the worst. The cumulative effect of the resultant public misconception was about to make wildlife management history.

Those who work professionally in the field of wildlife management have not had a tradition of having to publically defend their decisions, research programmes and goals. Understandably there had never been any need for defensive behaviour in Canada's marine research. Those who administered the research and who sat pondering population trends, statistics and herd health, were not trained media experts. They had not included in their schedules time to initiate press releases about their work. They were biologists, and functioned only in that capacity.

Therefore, although government personnel were disturbed by the effects of protest, there was no official move to counter it with information which would more closely approximate the truth. Both government scientists and university biologists were engaged in on-going programmes of research which were expected to take several to many years, and which would contribute to a growing wealth of published data on seals.

In 1967 the International Commission for the Northwest Atlantic Fisheries (ICNAF), which represented most of the countries fishing in the north and western Atlantic, also began gathering data on harp and hooded seals. There was at that time more research being conducted on seals and their immediate environmental relationships than had ever before been conducted on a single ecosystem. The irony was that although this had come about partly as a reaction to public interest in the seal hunt, the public itself was unaware that the concern it had expressed had caused any constructive response at all.

The message which Davies and others had been giving animal lovers was that the only desirable result of their protest would be an official ban on the hunting of seals. There was no protest suggestion that improvements in killing methods or population assessment would be acceptable to a well meaning international public.

Meanwhile, independent research into seal health, fertility factors, feeding patterns, natural mortality, parasites, behaviour of females with pups and behaviour of breeding animals, all continued without the knowledge or participation of the general public, which all this time was aware only of the annual protest.

The management goal of the '70s was to effect a steady, slow but measurable increase in the population of harp seals in Canada's waters. It was predicted that seal management would be proven possible with scientific supervision and adequate law enforcement to ensure that no cheating occurred in the harvest.

Public pressure to end the hunt continued, however, due to a consistent application of the protest message each spring. In response to this, the Canadian government decided to establish an indepdent advisory panel, the Committee on Seals and Sealing, to routinely observe and analyse the hunt and all aspects of seal research and management, so that outside independent expertise would have a bearing on ultimate management decisions. It was hoped that the formation of the Committee on Seals and Sealing would demonstrate

Canada's desire for a truly balanced perspective of the sealing issue, and would be an effective deterrent to protest charges that defense of sealing constituted a cover-up of the facts.

This Committee on Seals and Sealing was composed of non-government people who were completely independent of official government policy. Members included scientists, veterinarians, executive members of international humane societies, and a representative of the sealing industry. It was felt that this diverse and distinguished membership would indicate Canada's willingness to accept counsel from diverse perspectives.

The COSS would also be responsible for investigating the social and economic aspects of sealing in the Arctic and in the Maritimes since it was recognized that those who would be most affected by management decisions deserved a high priority consideration. The seals, after all, had been considered under the philosophy that they were a renewable, valuable resource, and those who benefited most directly from them should not be left out of the framework of management.

Even though Canada's programme of seal research had begun to gain international respect for its volume of data and its conservative quota policy, by 1972 the voices of protest were louder than ever. In the United States, Congress was hearing testimony by Davies and others which contradicted all claims of the humane slaughter of harp seals, and which questioned the advisability of continuation of the hunt at all on the grounds that the herds were in danger of depletion.

The Marine Mammal Protection Act of 1972 included a ban on the importation into the US of any of the products of marine mammals, including pelts, oil, meat or ivory. It stated that seals, whales, dolphins, walrus and other marine mammals were to be totally protected under the provisions of this Act, and that it would be illegal to kill or import them from that time on.

Although Canada objected to this legislation, the noisy strength of those who were doing the lobbying held sway over any other information. This route of legislative intervention in the free enterprise system was to become an important precedent for the future impact of the protest movement in its endless animal welfare campaigns.

In retrospect, there was justification for concern over some marine mammal species. The Marine Mammal Protection Act was a significant factor in influencing the rest of the world against the continued harvesting of some whales and dolphins.

The harp seal, however, had never been declared endangered by any scientific agency or organization, and the inclusion of this species under the general provisions of the Act was accomplished partly through ignorance in the Congress of population status, and partly because it would have been very difficult to counter the claims of Davies and the others without direct access to current research. These were anxious times for legislators who knew that the public feared for all marine mammals. They did not want to appear insensitive to public concern.

From that time on, the inclusion of the harp seal in the Marine Mammal Protection Act was held up by the protest movement as proof that its version of the issue was correct and not to be seriously questioned. The damage which this did to the credibility of Canada's marine research programmes and to her reputation in general was a bitter blow which many in the scientific community felt was entirely undeserved.

However, there was still an outlet for harp and hooded seal products in the Common Market nations of Europe, and the hunt continued as usual under the supervision of public and privately-sponsored research scientists, law enforcement agencies, and the Committee on Seals and Sealing.

Those organizations which sponsored successful spring protest demonstrations grew in number and in the skill with which they convinced the media that they had an "animal interest" story to tell. Greenpeace, the Fund For Animals, the IFAW, the Animal Protection Institute, and the Humane Society of the United States all achieved significant press coverage of their demonstrations and their goals. Cruelty and seal herd depletion to the brink of extinction were the major claims being made, and several movies "exposing the hunt" were produced for television.

This naturally kept the issue before the public and stimulated a continuing flow of letters to pro-sealing newspapers in Canada, the Prime Minister and the Minister of Fisheries and Oceans. Names and addresses were supplied to the public by more than one protest group.

This resulted in concerned members of the Canadian Parliament requesting more and more information about sealing policy from those who were in charge. Lengthy sessions were held during which scientists and bureaucrats alike were grilled about the wisdom of their goals and methods of management of this now unquestionably abundant resource.

A further cause for concern was the fact that Davies had taken his campaign of anti-hunt rhetoric to the European press, and was receiving wide support from animal lovers there, as well as in Great Britain. There was no stopping this flow of misinformation. Official protests, diplomat-to-diplomat or embassy-to-embassy, had no effect on international public opinion.

Since the effective message about the seal hunt reached those who vote, it was inevitable that anti-hunt legislation would be enacted. In 1981 the IFAW purchased ads in all the major newspapers of Europe, and Davies' own movies about the hunt were seen again and again.

By the spring of 1982 the European public had been tuned by the advertising campaigns of the IFAW to the extent that public demonstrations and massive petition drives denouncing the hunt were common. Greenpeace also stayed in the news, as its members marched before Canadian embassies and were seen releasing hundreds of white balloons with pup faces on them. The media referred to them all as "environmentalists."

March of '82 was the last time for the annual Greenpeace seal-spraying-for-the-media in the Gulf of St. Lawrence, and as usual it was an effective way of keeping up its high public profile. It was also good timing, for the Common Market (or European Economic Community — EEC) had just recommended to its member nations that there be a voluntary ban on the importation of all seal pup products into those countries for at least two years. In March of 1983 a directive was passed by the Council making the ban mandatory. This ban was to go into effect in October 1983 and last through October 1985 when the matter might be considered again.

Prior to this decision, there had been a great deal of diplomatic activity on the part of Canada to prevent the EEC ministers from going through with such a drastic action. Individual member nations were all contacted on the highest levels of government and the case for continuing the seal hunt was thoroughly presented.

The humane aspects of seal pup clubbing were again explained, defended, and endlessly discussed. Distortions of this method which had been given to the public by IFAW and other groups were countered. The humane training programme which sealers have to undergo before being licensed was also outlined. National leaders and their representatives to the EEC agreed that the hunt was as humane as possible, and probably more so than the slaughtering of domestic stock in the leading abatoirs of Europe.

However, none of this mattered because the voting public was not aware of any of it. There was no way to reach skeptical masses of people who would not have accepted another version of the cruelty issue anyway.

The same was true in the matter of seal herd strength and the management programs which had, through quotas and sophisticated counting and estimation techniques, thoroughly demonstrated that the seals were not in any danger.

However, the voting public believed otherwise, because for years most protest groups had been claiming that the seals were becoming endangered by the hunt. There had never been widespread public education on either the cruelty or the herd depletion claims, and in 1982 it was too late to begin such education.

The Common Market ban on seal pup products was a tremendous victory for the IFAW and the resultant strengthening of protest credibility also rubbed off on Greenpeace. The lobbying in the parliament, and the fact that Davies himself was appointed scientific advisor to the parliament on the seal issue are proof enough that the credit for the recommendations and the ban decision are his.

The final result meant that if men hunted seal pups in 1983, 1984 and 1985, there would be no market for them in Europe. Since seventy-five percent of all seal products would find no buyers, it was assumed that the commercial hunt would come to an end.

Seal pups were defined as whitecoated harps prior to their moult at the age of about eighteen days, and bluebacks, the hooded pups which kept that coloration for the first year. Both whitecoats and bluebacks had traditionally and commonly been taken with the club since they were always still on the ice in early spring and guns were neither appropriate nor necessary to their harvest.

The ban did not cover any older categories of seals, so that "beaters," "bedlamers," and old harps and hoods could legally be imported in the Common Market. However, the ban on pup products produced a reluctance on the part of buyers and speculators in sealskin futures to gamble on taking any sealskin. For this reason, the market for all sealskin pretty much disappeared.

Again, this background is relevant to a discussion of the management of these animals, because social factors which diminish the pattern of human predation, through either war or protest, mean a bottom line of significantly fewer seals being taken. More seals left in the environment mean that herds will be affected by other growth and mortality factors.

Although Davies correctly told the public that seals had been saved from slaughter by humans, he neglected to inform them of the implications of this. Therefore, it was widely believed that the seals would now live happily ever after, as long as hunting did not resume.

In 1982 the total allowable catch of harp seals was set by Canada at 186,000 animals, in a herd which was in accordance with scientific advice, estimated to number some two million and which was predicted to produce approximately 400,000 pups that year. The total actual take, including that by Norwegian vessels and Arctic peoples, was approximately 172,000 animals.

The usual combinations of weather, ice conditions and herd distribution across thousands of square miles always exert a certain unpredictable influence on access to harp seals. Some years the take equals or slightly exceeds the quota, and some years it falls far short. Over a ten year span, however, the total actual take has always been less than the total quota which was set.

A strictly controlled allowance is set each year for both the large ships which take whitecoats off the Front, and the catch by shore-based smaller boats. The harvest by Arctic natives is unrestricted since their access is so uncertain.

The annual quota is set on the basis of scientific advice, which is based on the counts of previous years, the tagging and plane surveys, and data collected on slaughtered animals. These quotas have consistently been set at the lower end of estimated replacement yields. This still allows a slow and predictable growth in the overall population. Some of the harvest always includes a number of animals which are one year old or more. Percentage-of-the-take penalties for taking mature females were levied in the late '70s until, by 1982, none could be taken legally.

The rationale for this is that, in general, young seals have a much higher mortality rate than do the mature ones, and a hunt which concentrates on them will have less effect on herd strength than would a hunt for breeding individuals.

Since harp seals do not breed until they are four or five years of age, the herd does not suffer significantly from the taking of a large proportion of those young, non-productive members. The fertile span of a female harp seal in a state of excellent nutrition may begin at age four and last through age twenty-five, or even longer. The life span of some has been recorded at thirty years. If food supplies are plentiful, and she is not weakened by parasites or disease, the harp seal female may have a pregnancy to term nearly every year of her life after she reaches sexual maturity.

Harp males in a state of excellent nutrition may attain sexual maturity at age four, but competition for females will include vicious fighting with older, dominant animals. In times of dense population, fighting wounds on sexually mature males are very common, and quite often debilitating. Infections and pain from repeated bite and slash wounds hamper the animal in obtaining enough fish to eat, and the healing process is slowed by his weakness and deteriorating condition.

Both males and females are affected by relatively poorer nutritional status in times when the herds are not depleted by human predation. This has been seen in the past when hunting was interrupted by the two world wars, and is expected to happen again if the hunt is significantly curtailed for some years due to the social factors outlined above.

The protest message about the hunting of any wildlife has always been that the population in question will "regulate itself" through natural controls once hunting stops. It is always claimed that these natural controls are less cruel and stressful for the individual animals than death by hunting could ever be.

In the case of seals, natural controls include a number of ways in which fertility is reduced as the herd becomes larger and seals congregate more densely in whelping and breeding areas. Mortality by natural means is also increased under these conditions, and the number of stillborn pups is always seen to increase, as is newborn pup mortality.

The net result, after a long period of greatly increased stress on each animal in the herd, is that fewer seals are born alive, fewer newborn seals reach one year of age, and fewer females reach sexual maturity, some never attaining it.

The basic reason for this lack of sexual development, low pup production and high overall mortality is simple. The seals in a large, dense herd are hungrier than those which live under less crowded conditions. Their bodies compute calories gained and calories spent, and the storage of surplus energy in the form of inner fat, and outer blubber, and react accordingly. Testicles and ovaries do not develop at the expense of those organs which control other bodily functions which must be nourished every day. The bottom line is that through chronic hunger, fertility of the herd is lessened and growth in numbers stops, then numbers decline. An environmental adjustment has been made, at a cost of discomfort and suffering for millions of animals.

One of the myths about wildlife which seems to persist is that there is a "balance of nature" which will prevail as long as "unnatural" events, such as heavy hunting by man, are not factors which affect the population. Those who have not had the opportunity to review the records of wildlife population changes over a number of years often believe that animals will, in the absence of hunting, eventually reach a stable peak population level and stay there. This is assumed to be the "carrying capacity" of the environment, a magic factor which can be reached if populations are just left alone long enough to achieve it.

Information similar to this has often been given the public by protest organizations which claim that the seal hunt is an unnatural factor in the environment and that herds were at a natural high level in the early 1800s before uncontrolled hunting severely diminished them. The usual claim is that herds at natural carrying capacity approached some ten million animals. The implication is that this number is one which most closely approaches the natural order of things in the Atlantic and that the present herd strength of two million animals is a sign that the harp seal is in extreme danger of extinction, or at the least has lost significant genetic variability.

The problem with this line of reasoning is that there is a basic assumption of a "natural" static carrying capacity. This is incorrect because there is no stasis in nature. Environmental factors are constantly shifting and there is no preordained balance or level of any of the many parts of an ecosystem. Not only do populations constantly change, but the very rates of change are always changing.

In the early 1800s, when seals were indisputably at a higher population level than they are today (although the ten million figure may be in doubt), they were enjoying an environment in which their prey, the several species of fish on which they regularly depend, was not being taken in any significant number by commercial fishermen.

In the twentieth century, the prey fish of harp and hooded seals are greatly affected by the activities of man. Not only are the same species taken by both seals and man, but those which are not taken by man are nevertheless affected by his selective fishing habits, and by his treatment of other predators, such as whales.

In today's complex marine ecosystem, in which the large fish still eat the smaller ones, even the krill is affected by man's presence and his decisions. It is the base of the entire food chain, vulnerable to pollution and eaten by small crustaceans, the fry of small fish, by the caplin, and even by the great humpback whale, which has managed to evolve a method of rounding it up into swirling pockets of food, and gulping it down; the humpback now consumes an estimated one half million metric tons of krill annually.

Man also takes the caplin which eat the krill. A temporary absence of one of its predators occasionally allows the krill to increase, to be eaten then in greater quantity by others, such as the fry and the tiny crustaceans. The whales, now protected almost entirely in the western Atlantic, also follow the krill and consume it to the detriment of those fish which depend upon it. Those fish lessen in number due to this diminished food supply and are therefore less available for the seals which gorge on them to excess each winter and spring. There is no end to the changes in supply and demand which are taking place as populations constantly adjust to one another.

Competition for food takes place not only between species, but within species, as whales compete with whales, seals with seals, and cod with cod for available energy supplies. There are no easy answers, as mutually workable balances are sought. The best that can be hoped for is continuation of all the members of the system in enough abundance that normal changes and catastrophic events will not wipe out whole species.

It is within this framework of understanding that those who seek to manage the seals, cod, herring, caplin, flounder, turbot, mackerel, shrimp and whales must operate.

To try to manage without complex knowledge is useless because there is then no predictable or beneficial result. Unfortunately, the benefits of twenty years of increasingly effective seal research and management, in conjunction with fish and whale management programmes, are now about to be erased by the impact of the protest movement. A poorly informed public has used its political power to bring about an end to the seal hunt and has therefore altered life conditions in the Atlantic. There is no reason to believe that this socially inspired management of seals will make their lives any easier.

VI

The Seal Saviours
Organizations of Protest

There are several thousand animal welfare organizations throughout the United States, Great Britain, and western Europe. The focus of their interests is very broad, ranging from prevention of cruelty in general to the concern for the survival of certain wildlife species. Some groups work to increase public interest in specific projects such as "saving" seals, dolphins or whales from any hunting, while others urge public support of general anti-hunting and anti-trapping legislation. Some advocate changes in the law in order to create legal standards for the housing and transportation of domestic stock, while others concentrate on the problems of pet overpopulation and pet abandonment.

Only a few such organizations have participated directly in the "Save the Seals" movement to end the Canadian and Norwegian hunting of harp and hooded seals. For each, the decision to become involved in this probably came about through a combination of circumstances.

The cynical observer may feel that the primary consideration for each was recognition of the economic potential to be realized from a focus on alleged cruelty to white seal pups. After the release of the Artek film and the new edition of England's book, Brian Davies' energetic IFAW publicity was recognized internationally as a fiscal success. Public acceptance of his message, media coverage of his protests, and the obvious affluence of the IFAW undoubtedly set an example for others and soon they adopted the harp seal as their own popular image booster and fund raiser.

The following are descriptions of some active "Save the Seal" campaigns carried on in the 1970s and '80s. Information was collected from the organizations themselves, their tax returns, the Assembly of Better Business Bureaus, and the *Encyclopedia of Associations*, 1984 edition.

The *Encyclopedia of Associations* can be found in any large library's reference section; it is annually updated. It lists organizations alphabetically within broad categories, gives their membership strength if this is reported, address of central office, presiding officer of the group, telephone number, and organizational purposes and activities. Animal rights groups are listed under "Social Welfare Organizations — Animal." The following is excerpted from the *Encyclopedia of Associations*' entry on the Animal Protection Institute of America.

The Animal Protection Institute of America

The Animal Protection Institute of America (API) is based in Sacramento, California. It was founded in 1968 and claims to have a membership of 100,000. Its purpose is to educate the public in the humane treatment of animals. Areas of specific concern have included "alleviation of whale, porpoise and harp seal slaughter," and informational progrmmes about threat of extinction of marine mammals have been a major activity.

API also advertises through television spots. A major focus of the group's "public education program" is an attempt to influence the viewer against all hunting and trapping. The purpose of the message is to cause a belief that all wildlife are endangered by human predation, and that all hunting causes cruel, unnecessary deaths. Television anti-hunting spots have included the harp seal pup face in a "fear of extinction" context.

A full colour poster issued by API includes the harp seal pup in a multi-species display which includes some creatures, such as the blue whale and the bald eagle, commonly known to be endangered. The poster is entitled "Wildlife Under Attack — Ten Of The Most Threatened North American Species."

A paragraph or two accompanies each picture and this text lists API's opinion of each creature's survival prospects, current population numbers, former population highs, and "How They Were Killed" or "How They Died," then "What You Can Do." One should be aware that not all of the creatures pictured are included on the United States endangered species list. Inclusion in this context of harp seals and bighorn sheep should be viewed with healthy skepticism for they are not recognized by any wildlife management authorities as "threatened" or "endangered." This "threatened" designation here is only API's opinion, although this is not necessarily evident to the uninformed casual poster reader.

Inclusion of the harp seal alongside the timber wolf, grizzly bear and American condor is a tactic which guarantees credibility for the implication that harps are themselves endangered. An ostensibly well-informed and well-intentioned humane organization has claimed concern for harp seals, whales, condors and grizzlies on the same poster. One assumes that harp seals must be endangered also. This "guilt by association" tactic has been used routinely by protest organizations and the general public tends to believe everything they claim or imply.

The inclusion of the bighorn sheep as one of the "threatened" and mention of it as a trophy object, serves to illustrate a common anti-hunting theme: that any species which is hunted is in danger. This reinforces the credibility of the poster claim that the "great whales" (represented by a picture of the endangered blue whale) as well as the harp seal, share this common danger. Thus any person who is impressed by the information on this poster is likely to agree with API or any other group which claims that harp seals should not be hunted at all for any reason

The poster also claims that harp seals die in a cruel manner due to the method of slaughter:

> Smashing the helpless infant with clubs until dead is a standard method of killing. Highly-organized huntsmen contended the clubbing would accomplish its aim at a single stroke. Often, it didn't. The baby seals are killed at the age of two or three weeks because the much-prized, snowy-white pelts soon change to a darker color....

This message also implies that there is a moral reason not to take the animals at a young age; they are "helpless." The poster states that "Since 1940, numbers are down by at least 9.2 million to less than 800,000. CANADIAN OFFICIALS DESCRIBE THIS SLAUGHTER AS A 'CULTURAL HERITAGE'." The poster goes on to state

> WHAT YOU CAN DO. U.S. prohibits import of these seal pelts now — other nations should do the same. Before the annual March hunt, lodge your protest with Canada and Norway. DISCOURAGE FUR-BUYING BY THE PEOPLES OF ALL NATIONS.

Thousands of good people have believed that their letters would have an effect on the hunt. However, neither Canada nor Norway has taken them seriously, and their decision-making has been unaffected by such letters.

Although most people have little awareness of the workings of other governments, it does seem particularly naive to expect that letters from foreign citizens should have any effect on another nation's wildlife management or trade policies. One might well question if API is equally naive.

Other material supplied by API to its supporters includes instructions on how to hold a successful seal hunt protest. Photographs of seal protesters carrying placards in English, French and German are featured alongside an article entitled "Fighting for the Harp Seal. Is It Worth It?" and "The Parties Responsible."

The photographs with the first article would lead one to believe that API supporters are protesting in France and Germany as well as in the United States and possibly Great Britain. The second piece urges concerned people to protest directly to those whom API targets as responsible for continuing the seal hunt. Addresses of government officials and major Canadian newspapers are given, and the message is clear; if API supporters write to these people, perhaps they will listen, and curtail or entirely end the hunt.

Additional instructions to sympathizers include "How To Hold a Seal Demonstration" and "Telling It On The Air — Key Points." This collection of information is all apparently intended to convince the reader that he or she can be a significant force in this worthwhile movement.

If this message is believed, and the reader follows through with a protest card or letter, then again, API has strengthened its credibility. The believer will be more apt to donate to API in the future, and to follow further suggestions about personal involvement.

The key to continued support may be in convincing one to become personally involved through protest or letter writing, and this, more than just the sending of support dollars, tends to reinforce the confidence which one feels in the organization. A person who has written a protest letter in response to any group's urging is entirely won over to the cause.

Any tactic which causes the personal involvement of a member of the public through a tangible response may be considered a successful marketing technique. The customer has "bought" the information if he or she then follows through and writes the letter, or organizes the protest.

The API success with its "customers" in the United States is demonstrated by its 100,000 membership and the fact that thousands of cards and letters have been sent by its supporters to the Canadian Department of Fisheries and Oceans and to the Halifax *Chronicle Herald*, a pro-sealing newspaper.

Although API supporters may be cheering the demise of the commercial seal hunt, they should realize that their own personal protest and the behaviour of their organization has had nothing to do with it. The Animal Protection Institute of America has had no impact on the seal management programme of Canada and has not been a factor in the institution of the ban on seal pup products in the European Common Market.

The API decision to join in the protest against the hunting of harp seals has undoubtedly enhanced its reputation at home as a humane-oriented organization. Its other anti-hunting and anti-trapping projects have probably been strengthened in credibility because of the association of API with seal protest.

There are factors other than animal rights activity which may influence the reputation of such an organization in the eyes of its public. For instance, the United States Internal Revenue Service no longer includes the Animal Protection Institute on its list of charitable organizations which are tax exempt. API lost this prized classification as of December 1983 and since that time American tax payers have been unable to list donations to the organization as deductions on their tax returns. Some potential donors may equate this loss of charitable status (and their loss of a tax deduction) with loss of respectability for API.

The IRS has two kinds of criteria which "charitable" organizations must continually meet in order to attain and keep the 501(c)(3) status: one is organizational and the other is operational. Thus both the structure of the organization and its behaviour must meet strict review standards. If the organization is found by the IRS to deviate from either, loss of status results. Since the API once had the status of charitable organization and then lost it, it may be fair to assume that its behaviour was a cause for the IRS to revoke its standing. If this is not the case, then a change in organizational structure caused the problem. Either way, the potential donor to any such organization might question the leadership directly about such a loss of charitable status in order to decide if a substantial donation or bequest is justified.

It is not known if the recent loss of charitable status has had any effect on public credibility of the Animal Protection Institute in the eyes of the public. Chances are good that relatively few supporters take the trouble to investigate any animal rights organization before making the ordinary donation for a yearly membership. Donation drives are based on emotional appeal to animal lovers, not on the relative merits of administrative procedure between one such group and another.

The Animal Protection Institute of America may be contacted by mail at: 5894 South Land Park Drive, P.O. Box 22505, Sacramento, California 95822, U.S.A.

Friends of Animals

Since Friends of Animals (FOA) is a well-known animal rights organization in the United States, it was questioned about its stand on the Canadian harp seal hunt. The organization is accordingly reported upon in this section.

Friends of Animals is an animal rights-humane organization with headquarters in New York City. President and founder is Alice Herrington who is also executive director of The Committee For Humane Legislation, the lobbying arm of Friends of Animals.

Friends of Animals was founded in 1957 and, according to the *Encyclopedia of Associations*, claims a membership of 100,000. It is concerned with influencing pet owners to curtail the breeding of unwanted dogs and cats. FOA offers "low-cost spay/neuter programs and assists animal shelters within the program." Friends also is "active in the banning of steel jaw, leg-hold traps and U.S. seal slaughter (in the Pribilof Islands), elimination of wildlife destruction, and boycott of furs."

A recent telephone conversation with a Friends of Animals representative also revealed that the organization has additional goals due to the wishes of "a major contributor" who (or which) was not named. The Friends representative stated that this major contributor had given FOA substantial funding on condition that it would actively lobby against any legislation which would act to enhance police powers. (Such a bill in the New York State Legislature concerning the power of the Law Enforcement Division of the State Conservation Department was, incidentally, *not* signed into law by the governor. Friends of Animals had recommended against the bill, as had the state's Division of the Budget.) A statement from the governor's office was made to the effect that the legislation would have been unnecessary and redundant and that the two negative recommendations had been noted. It is not known if one recommendation was more influential than the other.

One might legitimately ask why such a concern should be a part of animal rights interest. The answer may lie in the size of the donation to Friends. The "major contributor" was not named by the informant, who was not aware that she was speaking to a member of the state police force.

Ordinary donors to Friends or those who otherwise support the organization might wish to inquire into this matter themselves if they feel that this is indeed a peculiar and socially-harmful result of Friends of Animals' acceptance of funding. Of course, the answer to the question may lie in a philosophy which states that a strong police force diminishes the rights of animals, although this is patently absurd. Or one might infer that the organization cares little about the conditions of funding, as long as funding in substantial amounts does come through.

Alice Herrington wrote this author that Canadian seal slaughter has not been a topic pursued by FOA and that the International Fund For Animal Welfare could be consulted for information on the Canadian issue. Herrington also commented that journalists and photographers do not have real freedom of access to the Canadian hunt, due to Canadian suppression of the information they would obtain. It is not known if she or a Friend of Animals representative ever applied for access to the hunt, or had been refused such access.

Friends of Animals has not spent any effort on raising funds to protest the harp seal hunt. The basis for this as an economic or philosophical decision is not known. Perhaps Herrington felt personally that the Pribilof hunt should be given priority by Americans or that the problems of observing the Canadian hunt and reporting adequately on it were just too great.

Friends of Animals is listed by the U.S. Internal Revenue Service as a charitable, tax exempt organization.

The Philanthropic Advisory Service of the Council of Better Business Bureaus, Inc., was queried about its files on Friends of Animals. The Council replied that Friends had been repeatedly asked to respond to the Better Business Bureau regarding its accounting and organizational practices, but as of December 1982 Friends of Animals had not done so. Therefore, the Better Business Bureau could not determine if Friends of Animals meets any provisions of its own standards for charitable organizations.

Since Friends of Animals has had nothing to do with protesting the Canadian harp seal hunt, the organization has had no effect on public opinion regarding this issue and has had no impact on the demise of the commercial seal hunt in Canada.

Friends of Animals may be contacted by mail at: 11 West 60th Street, New York, New York 10023, U.S.A.

The Fund For Animals

The *Encylopedia of Associations* lists the Fund For Animals (founded 1967, president Cleveland Amory) as a 200,000-member organization. The Fund is based in New York City and, according to the *Encyclopedia*, its reasons for being are: "To protect wildlife, save endangered species, and fight cruelty to animals, both domestic and wild, by means of legal action, direct activism, public education, and lobbying."

A major effort has been to publicize the Fund's opinion of the Canadian seal hunt. Cleveland Amory himself went to the ice in Brian Davies' helicopter and, on at least one occasion, walked about out there with Tom Hughes, executive vice-president of the Ontario Humane Society and a member of the Canadian Committee on Seals and Sealing. Thus Mr. Amory had the opportunity to observe the seals in their undisturbed state, although it is not clear whether he actually observed normal hunting of seal pups himself.

Amory also went to the ice in an old ship purchased by the Fund for that purpose in the late 1970s. His account of the seals' behaviour and that of his band of "seal saviors" was subsequently published in *Good Housekeeping* magazine. "Lets Save The Seals" includes an Associated Press photograph of Amory holding and hugging a heavy seal pup, which he claims to have taken on board for the occasion. He stated that its mother "seemed to know we meant him no harm." (Such a large pup had probably already been abandoned.) Amory's crew set about to spray red dye on the whitecoated pups, allegedly under cover of darkness. He claimed that more than 1,000 were covered. The point of the dye application was to ruin the pelts so that those individuals would be spared by seal hunters. All those spraying seals were subsequently arrested by Canadian authorities.

This type of behaviour on Amory's part has been termed media grandstanding by many in the field of wildlife management, and Amory has been known to carry out other media-worthy events in the United States in order to "save" whitetail deer, jackrabbits, or burros which were being hunted for reasons of population control.

The *Good Housekeeping* acceptance of Amory's seal story was certainly a victory for his credibility as a major seal protester, and must have had a positive effect on donations to the Fund. The short (one page) piece with photograph was later reprinted for general distribution to Fund membership in their newsletter. That he had an Associated Press photographer along on the ship attests to his promised appeal.

Although Amory's anti-hunting behaviour and its impact in the United States has been well documented in "predator" magazines such as *Outdoor Life* and others, these criticisms, no matter how well written or blunt, serve only to firm up the opinions of that part of the population which supports hunting and buys the magazine. Thus, that small segment of the general public (200,000) which routinely supports the Fund For Animals has not been exposed to any meaningful criticism of Amory's philosophy or behaviour since few would be caught dead reading a magazine written for, by, or about hunters.

This writer asked Cleveland Amory about his personal involvement in the seal hunt protest and about the Fund's stand on the hunting of harp seals in a letter which he decided to answer with a telephone interview in June 1982. He stated that his organization shall "always be active" in seal hunt protest and that he supports both the IFAW and Greenpeace as philosophical partners in this effort.

Amory credits Brian Davies as being the real leader in the fight to end the hunting of harp seals, although in a later conversation he stated that the IFAW fish boycott would "not be necessary" and would probably not be put into effect since the Common Market ban on imports of pup pelts would put such a damper on hunting. Perhaps Mr. Amory rather naively believed that Davies would drop seal protest due to his victory in Europe.

When asked if he felt the harp seal population was in danger due to hunting, Mr. Amory stated "That is not an issue with me. Cruelty is the issue." He went on to say, "This clubbing is the cruelest single mass event that goes on in the entire world. The baby beside its mother makes it worse."

Use of the cruelty issue appears to be the sole selling point for the Fund For Animals' harp seal protest. The Fund at one time encouraged its supporters to send postcards to Canadian tourism officials stating that they would not visit Canada until the seal hunt was stopped. Amory himself noted that this effort became meaningless when the exchange rate became favourable for Americans. Two hundred thousand postcards would not necessarily mean a loss of 200,000 tourists, however, and no one in official Canada seemed to pay any attention to the mail onslaught. This "involvement of the constituency" tactic of card- and letter-writing undoubtedly strengthened and confirmed the loyalty of those who believed Amory's message.

Another tactic used by Cleveland Amory to influence public opinion on the seal hunt has been to encourage the media to cover celebrity endorsement of his views. Amory recalled this as an impressive demonstration which he had staged during a 1978 press conference, and listed Mary Tyler Moore, Henry Fonda, Burgess Meredith, Burl Ives, and Loretta Switt as "helping out" in the effort to protest the seal hunt.

This tactic of celebrity endorsement, of course, has been used to sell everything from beer and cigarettes to floor tiles, and one must assume that it is equally effective in selling the story of cruelty to seals. One must remember that the good people who have given their endorsement to any "save the seals" effort have probably been genuinely impressed with the information which they have been given on the matter and have felt that to endorse such an "obviously humane" effort to end the hunt could not hurt their public image or upset their own feelings of self-worth. It's a no-lose issue to stand behind, if one has no knowledge of any facts other than those put forth by the side of protest.

Amory himself was billed as a celebrity in *The Hunt Without Pity*, a film produced by the IFAW during the 1970s. He was billed as a famous author and humane writer, and in the film spoke of "the cruel effect" which the pup hunt has on mother seals, and stated positively that "there is no use of the meat." Those thousands of Newfoundlanders who routinely eat seal would be amazed that he could say such a thing. Since he had done his protesting in the Gulf of St. Lawrence and had not spent any research time in Newfoundland, apparently Mr. Amory was unaware of this use of seals by the hunters and their families, and the fact that ship loads of flippers from the whitecoat hunt went to Newfoundland each season.

The Fund For Animals has charitable organization status in the United States and the IRS 501(c)(3) rating means that American tax payers can deduct their donations to the Fund on their tax returns. The Better Business Bureau states that the Fund For Animals meets its own standards for proper charitable solicitations.

The Fund For Animals has not had any significant impact on those developments which have ended the commercial hunt. Fund activists have not been influential in European Common Market nations. Fund supporters have not influenced Canada in any way to change its seal management policies. Therefore it is reasonable to consider that the money which Fund supporters have sent on behalf of seal protest has had no real effect on seals, individually or as a species.

The Fund For Animals may be contacted by mail at: 140 West 57th Street, New York, New York 10019, U.S.A.

The Humane Society of the United States

The *Encyclopedia of Associations*' entry for the Humane Society of the United States is quite lengthy compared to that for many other organizations. This is a function of what was submitted to the editor by the organization rather than an indication that HSUS is necessarily more active or effective in its field or as a fund raiser than are those organizations for which the entries are much more brief.

HSUS was founded in 1954 and claims a "constituency" of 200,000. Its interests cover ever conceivable humane concern for wild and domestic animals. In general HSUS is anti-hunting and anti-trapping. It tries to exert significant influence on the enforcement of existing humane legislation and attempts to lobby for new and ever more restrictive laws governing the use of any wildlife as a resource.

The Humane Society of the United States claims to promote kind and responsible treatment for all domestic pets and stock, and conducts a major humane education effort in schools.

Although the Humane Society of the United States has appeared to conduct a major campaign against the hunting of harp seals, the effect of the organization on the demise of the commercial hunt has been nil. None of the efforts to raise public consciousness about the hunting of harp seals as a moral, humane or ecological issue has resulted in any specific action taken by Canada, Norway, or the Common Market nations of Europe.

However, it would be incorrect to say that HSUS had merely appeared to use the issue as a domestic fund raiser without putting forth any international effort at all. The organization did send at least one representative to the Convention for International Trade in Endangered Species (CITES) when it met in Botswana in 1983. At that meeting, member nation West Germany attempted to have harp and hooded seals added to one of the endangered lists. HSUS knew about this effort beforehand and asked its constituents to write President Reagan asking him to instruct United States scientists to go along with this move. HSUS was very active at the meeting, lobbying delegates from many nations to vote to declare these seals endangered.

The effort failed completely. The seals were not declared endangered, as scientists from many nations knew that it would be inappropriate, for biological reasons, to do so. CITES member nations send their scientists to each meeting to discuss certain issues pertaining to the population status of their species of concern. These discussions are based on genuine scientific research and on predictions of population changes in each species.

A scientific delegation to the CITES meeting in Botswana told this writer that there were more protest organization representatives at the convention than legitimate delegates. Although the non-scientists were not allowed to vote, they could sit in on plenary sessions and could state their views. They also spent a great deal of time and effort after-hours in attempts to wine-and-dine delegates in order to influence their votes on specific issues.

One might ask if this use of donors' money on behalf of seals or any other species could ever reflect the best interests of donors, since it supported no *scientific* basis for vital decision-making. The impact of protest organizations on the delegates to the CITES meetings is apparently not one of academic persuasion (after all, they are not full-time research scientists), but one of social and political pressure. Such pressure became so overwhelming at Botswana in 1983 that an unprecedented secret vote on the seal issue was adopted since some nations complained about the amount of pressure they had received.

Those who have donated to HSUS and similar organizations in the belief that they were supporting a sound *scientific* approach to the welfare of certain species, should be aware of the total lack of scientifically justifiable impact which such organizations can exert.

Each donor to a protest organization who is impressed by the fact that the group sent one or more representatives to a CITES convention should know how to ask about the justification for that expenditure as an action useful to the welfare of the species. Additional questions about the role of one's protest organization might be asked of one's governmental representatives to CITES. A much different picture of what actually goes on might emerge.

Each organization which sends someone to a CITES meeting can, at the least, claim beforehand that a representative will go to try and convince the scientists of the world that the wishes of protest are appropriate to solving the problems of the species. Newsletters to memberships of large humane organizations announce such action and these announcements are impressive tools with which to reinforce constituency beliefs and values and aid those groups' credibility.

An animal rights supporter hears that his group is going to CITES to fight to have his favoured species added to an internationally recognized list in order to curb hunting. This is the whole point of protest and one is proud of the organization for making this serious, potentially meaningful attempt. Again, the ordinary donor, member, or constituent feels that his financial support has been a significant part of a worthwhile effort. His opinions concerning the welfare of the favoured animal have been taken to an international forum by his trusted organization. His organization must be powerful indeed to command an international audience for these opinions. The back-home credibility of the organization, and continuation of financial support from the constituency, is assured.

From the perspective of an organization's administrators, therefore, it is important to send representatives to international conventions which discuss the legal status of favoured species. This always enhances the image of the organization in the eyes of its constituency. The organization can not lose by this expenditure, for it can always say "We tried our best on behalf of the animals. Next time we will be there again, and we'll try harder."

HSUS has portrayed the harp seal pup hunt as cruel, ecologically harmful, and economically unnecessary to those who conduct it. Recent publications and a bumper sticker reinforce and magnify the feelings of supporters that HSUS is doing everything possible to stop seal hunting, not only in Canada but in the Pribilof Islands as well. The two entirely different seal hunts make use of clubs as the killing device. Both hunts have been declared inhumane by HSUS.

Publications sent to the constituency include coloured photographs of seals being clubbed in the Pribilofs and of an HSUS staff member holding a white pup on the ice. The pup is squinting its eyes shut, in a grimace typical of that seen any time one is encircled by the arms of a human; it has gone into the deep diving reflex, perhaps because the human had touched it on the forehead so it would hold still for the picture. A person who was unaware of this reaction would perhaps think that the pup was exhibiting an expression of contentment, or even smiling. Any attempt to hold a pup without causing the deep diving reflex would result in a struggling, biting animal which would not be a very attractive or safe photographic subject. Sue Pressman, HSUS director of wildlife protection, must have been at least minimally familiar with this behaviour of the species in order to have accomplished such a pose.

HSUS in 1980 was encouraging its supporters to have faith in the credibility of the campaign to end the hunt by enlisting their personal action. Of course, this consisted of letter writing; the name and address of the current Canadian ambassador to the United States was given. It was urged that one "Tell the Ambassador you abhor the clubbing.

Ask him to respect the majority opinion of his own citizens as well as that of the rest of the world by stopping the slaughter now!"

However, diplomats to foreign countries do not have any control over their nation's internal wildlife management policies. This is not their line of work, nor their area of expertise. It is likely that the HSUS advice to potential letter writers, futile though it was, was never questioned by them. A sincere faith in the advice of each protest organization has been repeatedly demonstrated by a small to moderate yearly flow of such letters to the office of each Canadian ambassador since the protests began in the late 1960s.

Prior to 1983 HSUS identified some other villains for its supporters: the ambassadors to Norway and West Germany were also pinpointed as culpable in promoting seal hunting since each country derived a profit from it. Both embassies heard protest from the people who believed in the claims of HSUS.

A boxed-in paragraph in the 1980 Close-UP Report of The HSUS is typical of the usual request for help in spreading this message:

> Help the HSUS. You can help The HSUS continue to oppose the cruel clubbing of baby seals as well as our other anti-cruelty activities by sending a contribution to The HSUS. Your gift will greatly assist us in furthering our determination to create a society in which ALL animals are protected from cruelty and abuse.

The newsletter also contained the information that HSUS would send Sue Pressman on a press-contact tour of the United States, and that "fact sheets" about the hunt would be mailed to newspapers and radio and television stations across the country. "It is our intention to contradict the Canadian claims in every forum possible. Pressman's one-to-one contact with the press will help gain a more rational view of the subject than the Canadian PR agency tour of the US."

It is without question that the HSUS campaign to denounce the hunting of seals anywhere has always been a rational one, from the standpoint of goal achievement and information dissemination. The goal achieved has been an apparently perfect credibility for HSUS with its supporters. The information given out has been that all hunting is unjustified and immoral and that seal hunting is an especially good example of unjustified resource use.

Although HSUS announced that its policy statements on the seal hunt would be sent to media outlets across the country, it did not go so far as to claim that these might be run as news or public service announcements. It is suspected that most are paid advertisements and that the constituency, if polled, would be behind the expenditure one hundred percent.

The Humane Society of the United States is tax exempt under section 501(c)(3) of the Internal Revenue Service regulations for such charities and meets the Council of Better Business Bureau standards for charitable organizations.

HSUS appeals to a sense of compassion and righteousness in American animal lovers who are not, at the same time, sportsmen or sportswomen. The latter do not appreciate HSUS slogans and continued attempts to curtail their taking of game. They would be wise to keep track of the effect that HSUS may be having on the behaviour of their state and national legislators. Substantial contributions to the election campaigns of legislators are a political fact of life in any power game, and those Americans who hunt and trap should make it their business to know if their representatives ever accept contributions from HSUS or any other animal rights organization. Such a debt on the part of their lawmakers can only bode ill for their own interests.

Those who hunt and trap may fail to realize that they are such a large part of the electorate because their views have received less media attention than those of the many animal rights advocates. A simple tally of the members or supporters of all the animal rights groups in the country can be made with the aid of the *Encyclopedia of Associations*. This clearly indicates that the animal rights total falls far short of the number of those who buy hunting and trapping licenses.

The relative political power of the two opposed interest groups, however, may be weighed in terms of organized action to influence legislation. Hunters and trappers have the National Rifle Association, the National Wildlife Federation, the Wildlife Legislative Fund and other lobby groups to serve their interests, but the struggle to influence those who write meaningful legislation is unending. This is part of the democratic process, and the dynamics of the system demand constant attention.

Organizations such as HSUS, Friends of Animals, the IFAW, and the Fund For Animals are proud of their record in state and local governments. Animal rights organization achievements on the federal level include a voice in the passage of the Marine Mammal Protection Act of 1972. This legislation prevents the importation of the products of any marine mammals into the United States regardless of whether or not the creatures are endangered or a necessary part of the economies of other countries. HSUS is one organization which points with pride to its involvement in the enactment of the Marine Mammal

Protection Act. Its message is one of power, not one of biological justification for the legislation. The demonstration of such power, or the organization's claim that it has exerted it, is enough to satisfy many a potential donor whose values are reinforced by the rhetoric of the organization.

Perhaps, for such a person the claim of power on the part of the organization is enough to justify that organization's actions. The old cliche "might makes right" is operative here, and the tradition of animal rights advocates has been to force their value system on the general population as often as possible. No regard for the human rights of other cultures, or sense of tolerance for other value systems, is ever evident. The lack of regard for other peoples' life styles and economic networks is similar to that of a religious missionary fervour and animal rights groups behave as though faith and credibility alone are enough to justify the means to the end.

The HSUS slogan, "Club Sandwiches, Not Seals," absurd as it is, adorns sweatshirts of followers such as Vermont Congressman Jim Jeffords, who posed with other HSUS supporters on the steps of the Capitol to celebrate the first National Day of the Seal, March 1, 1982. Congressman Jeffords had not been able to talk the entire House of Representatives into endorsing Congressional Resolution No. 236, which would have made the Day of the Seal an official national observance. But the resolution did attract more than one hundred co-sponsors. However, according to HSUS, a companion resolution introduced by Senator Lowell Weicker (Connecticut) was amended by the full Senate which "without objection, declared March 1, 1983 National Day of the Seal."

A rally was held on the steps of the Capitol and HSUS official Patricia Forkan told the crowd, "We are here to celebrate seals, not club them." Others who addressed the crowd that day were Congressman Jeffords and Congresswoman Claudine Schneider of Rhode Island. Jeffords "movingly described his trip to the Canadian seal hunt as a member of the congressional delegation in the early 1970s. 'As I stood there in the pristine arctic beauty, I was shocked at the brutal killing to obtain seal skins,' he said."

This writer talked with a Newfoundlander who claimed he had met Jeffords on that day out on the ice. The Newfoundlander said Jeffords asked him, "What's going on here?" and he replied that it was a normal seal hunt. Jeffords then allegedly remarked that it looked very much like the Vermont practice of hog slaughtering, no better and no worse.

It probably doesn't matter what Jeffords or any other Congressional representative ever said to a Newfoundlander. Their statements since have been much more meaningful in accomplishing the goals of the anti-hunting establishment.

The use by animal rights organizations of celebrity endorsements is a tactic which sells a lot of credibility. The use of celebrities who are also legislators is a double bonus suggesting real accomplishment, and the only way to ensure the continuing power of this minority.

The Humane Society of the United States may be contacted at: National Headquarters, 2100 L Street N.W., Washington, D.C. 20037, U.S.A.

The Seal Rescue Fund

The Seal Rescue Fund is operated by the Center For Environmental Education in Washington, D.C. This Center also sponsors and operates the Sea Turtle Rescue Fund and the Whale Protection Fund. The Seal Rescue Fund was begun in 1980 and has supported itself with direct mail fund raising campaigns.

According to the *Encyclopedia of Associations*, the Seal Rescue Fund has attracted 63,000 members. The aim of the organization is "to maintain the seal population worldwide and protect it from potentially harmful human activities." The Seal Rescue Fund is primarily concerned with threats to whole species, with population depletion and extinction as its major concern.

Actual work conducted by this organization is claimed to consist of using "realistic, rather than idealistic, methods to achieve goals. Conducts research; analyzes and applies data. Engages in direct conservation action; cooperates with governmental agencies and legislators...conducts educational programs and research on seal habitat, population, and management. Current projects include 'work' on the Interim Convention of the Conservation of North Pacific Fur Seals, the Hawaiian Monk Seal, seal/fishery interactions, and habitat degradation. Sponsors the International Day of the Seal."

The Canadian harp seal issue is of interest to the Seal Rescue Fund because of the Fund's "desire to maintain the ecosystem integrity and biological diversity of the seas." This statement apparently indicates that the SRF feels the taking of harp seals under the recent quota system is a threat to their numbers, and thus to their genetic diversity. A letter to this writer from the Seal Rescue Fund (June 29, 1982) states that hunting as it has been conducted has resulted in numbers "far below its original abundance and its future is uncertain."

To support this contention, the Fund referred to a study of harp seal population dynamics by a Dr. J.R. Beddington of England. Beddington's original work claimed that the harp seal population was "actually declining, rather than increasing at a 5% growth rate, as NAFO [North Atlantic Fisheries Organization] contends."

The Seal Rescue Fund, noting that NAFO had dismissed Beddington's work, claimed to have contacted "an independent population dynamics expert" to review his paper in order to come up with "the best scientific base possible for any proper marine mammal management scheme." The goal was "to permit prediction of effect, not only on the target species, but on its whole ecosystem."

As it turned out, the expert under contract was Professor Daniel Goodman, a reputable biological statistician from Scripps Institution of Oceanography and the Biology Department of Montana State University at Bozeman. Dr. Goodman subsequently served on the ICES Panel, along with Dr. Beddington and sixteen other scientists.

The panel's official published conclusions on the status of harp seal population figures and pup production have since been interpreted in various ways due to a certain unfortunate but unavoidable ambiguity in the original paper. Figures given, however, are as follows:

Year	Pups (thousands)	*Population Age 1 year + (millions)*
Late 1960s	320 – 420	1.2 – 1.6
1977-80	380 – 500	1.5 – 2.0

The panel stated that the data showed a "central tendency," which means that "estimates close to the calculated value are more likely than are estimates far from it." Although unknown biases could be working to affect the data in opposite directions, the extreme values "are not considered as likely as intermediate values,....the Working Group did not attempt to quantify this tendency in probability form."

The panel, or "Working Group," concluded that "the pup production in 1977-80 and 1+ population was likely to have been larger than the late 1960s pup production and 1+ population, but the possibility of no increase or a slight decline is not negligible."

This last sentence is the reason the "contracted expert" (Dr. Goodman) of the Seal Rescue Fund has objected to the "unwarrantedly optimistic" press release of the Canadian Department of Fisheries and Oceans. This release (December 1982) gave the above pup population figures, and stated that the ranges "allowed for all possible sources of error in estimation." While this is a true statement, the release did not mention that a slight decline or a slight increase in pup production or general (aged one year plus) population was a chance that was "not negligible." Anyone can quibble over how significant the terms "likely" and "not negligible" are. No scientific paper is apt to state any interpretation of data in unequivocal terms because the nature of most data is that there is room for statistical variance in the "conclusions." The general public always wants a simple, straight-forward answer, and press releases intended for the public reflect, of necessity, this "simplicity" of interpretation of almost anything discussed.

One might ask why Canada should be damned for failing to say that there was a "chance" that either pup production or adult harp population could *either* have increased or decreased slightly more than the numerical values given. The "chance" was "not negligible," meaning that it has to be taken into account in a serious discussion among scientists, yet there has been no statistical probability assigned to it.

Harp pup production and population had "likely" increased over time; this had not been quantified as a probabilty by the working group and the author(s) of the release chose not to discuss or enlarge upon that. The release did admit that the panel's findings on hooded seals showed data was insufficient to say just what their status was, although a decline in their numbers was not evident. In response to this, the Canadian quota on hoods was then reduced by 3,000 animals for the coming year.

The expert who had been hired by the Seal Rescue Fund to study and discuss harp seal population dynamics is a reputable statistician whose "contractual status" with a fund raising environmental protest organization may or may not have been known by his fellow members of the ICES panel. His private arrangements should not be of consequence in any assessment of his performance as a professional analyst. It would not be justifiable to suggest that he might have been predisposed to criticize the panel for concluding that harp seals were perhaps not depleted at all by recent harvest levels.

If, however, he had any serious reservations about the wording of the conclusions of the panel, these should have been mentioned by him while it was in session, and any problems would have had to be resolved to the satisfaction of everyone, including himself. The panel report does not mention that Goodman had any specific statistically based objection to the wording or figures of the "Trends in pup production and population size" section.

It seems fair to conclude that the panel members agreed on the wording since ICES reports are composed through such a process of concensus. Goodman's paper, which he had presented to the panel, was later (December 1982) delivered to the Seal Rescue Fund. It is a criticism of Canadian seal management (based on his own prior interpretation of the data) and of the panel's procedures. It also criticizes the Canadian press release on the conclusions of the panel as "unwarrantedly optimistic."

Dr. Goodman's fellow scientists may wish to study his postscript conclusions and interpretations and decide for themselves if they call for further discussion. His paper is in the possession of the office of the Seal Rescue Fund, Center for Environmental Education. Since the Fund claims to be supportive of the scientific community, it would probably not object to distribution of this entire paper for general discussion.

Although the Fund expresses interest in building knowledge about all the parts of an ecosystem, it appears to have rejected all the NAFO scientific studies on the harp seal as unworthy of mention, even though they include discussion of seal and prey (fish and shrimp) interactions. NAFO studies have upheld the conclusions of the ICES Panel, perhaps in part because nothing more definitive has been produced. Predictably, the SRF rejects the conclusions of the CITES Convention in Botswana on the seal issue.

One might fairly conclude that the Seal Rescue Fund is dissatisfied with the opinions of the scientific community on the topic of the status of the western Atlantic harp seal. In fact, the Fund has stated (June 1982 letter to this writer) that the "deluge of public opposition to the harp seal hunt" should have convinced Canada and the North Atlantic Fisheries Organization to take action to eliminate it. This is not a scientifically justified approach to ecosystem management, although it is apparently attractive to the general public.

It is doubtful that public education is actually enriched by this attitude. Popular opinion, encouraged by an anti-hunting protectionist bias, is not a meaningful substitute for scientific investigation when making management decisions. It appears that the Seal Rescue Fund has in this case used the public's already formed opinion of the harp seal hunt to further its own goal. The "Centre for Environmental Education's goal is to educate the public and encourage them to take initiative in conservation policies while retaining our scientific credibility and political respect in the policy making arena." It is not known if this latter goal has been met.

The intent of this discussion has been to acquaint the reader with the Center For Environmental Education and its Seal Rescue Fund in a brief and fair way as far as it has been concerned with the fate of the harp seals. The CEE has not conducted protests of the scope or kind engineered by other "conservation" organizations, although it has used letter- or card-writing to some extent in its attempts to demonstrate public opinion in some matters.

The harp seal issue, however, has never been presented by the Center or the Seal Rescue Fund in as theatrical a manner as has been employed by Greenpeace, the Animal Protection Institute, the International Fund For Animal Welfare, the Fund For Animals, or the Humane Society of the United States.

The killing of harp seal pups, however, has been described as "baby killing" in Fund literature, and this has undoubtedly contributed to the public's perception of the Fund's concern. Thus, the taking of infant animals as a moral issue is alluded to, but the biological impact of selective hunting for immature animals instead of for breeding stock, has not been a part of Fund educational literature.

It would seem that the Fund has used the public's low opinion of the harvesting of "babies" in the same manner as has Greenpeace and other groups, which at the same time stress "the humane issue" as a central theme. The Seal Rescue Fund does not stress the humane killing issue in the case of harp seals, and has felt that for the group's campaigns, cruelty as a major topic has overshadowed the more significant ecological impact of taking large numbers of the animals. The ostensible concern of the SRF has been for species welfare entirely and, as pointed out above, the organization has implied that the killing of pups is somehow "wrong" but has not made a big point of this. Perhaps it was felt that this would not offend an already "educated" public on the humane part of the harp seal issue, and that it would not hurt to remind people that this is the same seal as has been shown them in other literature.

It would appear that the Center For Environmental Education has attempted to market its appeals to a more highly educated, more environmentally aware, segment of the public than some other groups have concentrated upon. This is the opposite of the type of approach used by Cleveland Amory who apparently wishes to impress the average "humane woman," or of the Humane Society of the United States or the Animal Protection Institute, which appeal in general to a very pet-oriented public. It may reach fewer people over-all, but gain a higher response rate from a more selective mailing list.

The Center For Environmental Education, and the Seal Rescue Fund, have 501(c)(3) status with the Internal Revenue Service, but for some reason only twenty percent of a taxpayer's donation may be used as a deduction.

The Center For Environmental Education does meet the Council of Better Business Bureau's standards for charitable solicitations. The Better Business Bureau's report on file for 1980-81 indicated that the Center For Environmental Education stated that its "educational programs" were carried out in conjunction with direct mail fund raising.

The separation of expenses is a complicated subject; in 1982 the Better Business Bureau disclosed that the Center spent seventy-seven percent of its first year's donations on direct mail costs, although it (the CEE) considered seventy-three percent of those costs to be in "public education program expenses." In other words, one could be educated at the same time as s/he could also read a plea for a donation somewhere in the information packet. This is not unusual. All charitable or protest organizations which depend on the public for voluntary funding take the opportunity to ask for money somewhere in their literature, amongst the "program services" which they conduct.

The Center For Environmental Education may be contacted at: 624 Ninth Street N.W., Washington, D.C. 20001, U.S.A. (telephone 202-737-3600).

Greenpeace

Greenpeace is an international organization which originated in Canada and has chapters in many western world countries. Most of these appear to support the philosophy and projects of Greenpeace International and apparently send dues in to the mother organization. Greenpeace declines to call itself an animal rights group.

Greenpeace USA is listed as an "environmental" organization in the *Encyclopedia of Associations*. A favourite slogan is "We are Environmentalists."

Greenpeace protest of the harp seal hunt has received as much media notice as the attention it has given its anti-whaling endeavours for "ecological," "humane" and, probably, fund raising reasons.

The purposes of Greenpeace USA fall within the general range of those of the international organization, in that it claims to be made up of "conservationists" who conduct active though non-violent protest of activities which they consider ecologically harmful. A statement of purpose appearing in a Greenpeace USA brochure issued in 1984 is as follows:

> In the broadest terms, Greenpeace is working towards the achievement of an ecological consciousness, whereby humanity recognizes that it is both totally dependent upon, and an integral part of, the global ecosystem. A few of our immediate goals are: A world-wide end to the commercial slaughter of whales, dolphins and seals. A halt to the proliferation of nuclear weapons and nuclear wastes. The protection of endangered species. Clean air regulations resulting in an end to acid rain. An international ban on the dumping of hazardous wastes into the oceans.

It is not known if these were given in order of importance, according to Greenpeace, or if they were listed randomly as examples of Greenpeace concern.

The Greenpeace "way of doing things" is explained as direct action to prevent ecological disasters:

> Our best-known tactic is direct, non-violent action: High speed, inflatable craft interferring with whaling operations; people stopping sealing ships from breaking through pack ice; skindivers freeing dolphins; and protestors scaling smoke stacks. The main purpose of such tactics is to draw public attention to critical environmental issues, so that the full weight of public opinion can be brought to bear on our decision-makers.

Greenpeace takes credit for bringing whaling as an ecological "crime" to the attention of millions of people "who only ten years ago...were unaware that modern whaling had virtually eliminated many species of great whales." Greenpeace claims that its intervention before the International Whaling Commission resulted in the institution of bans on the commercial hunting of all whales by 1986. Greenpeace claims credit for a worldwide ban on the hunting of sperm whales, and for a ban on factory ship whaling in general, covering most of the world.

The background of the International Whaling Commission as a body of research scientists who make decisions about the use and protection of whales as a resource is not discussed in Greenpeace literature. The Greenpeace claim is that the IWC has been extensively influenced by this body of non-scientists, because of the massive public support enjoyed by Greenpeace on the anti-whaling issue. A member of the public reading the brochure would have no idea if these claims of influence were justified, and would perhaps be disinclined to question them. They are given as facts under "ACCOMPLISHMENTS."

The same sort of information is given under Greenpeace accomplishments for seals, dolphins, toxics, and nuclear issues. One might note that Greenpeace claims credit for the European Economic Community's ban on importation of "baby" seal products into Common

Market countries, thus "resulting in a collapse of traditional seal markets and an 80% reduction of 1983's slaughter." In fairness, the small print at the top of the page under "ACCOMPLISHMENTS" does state "by working independently and in parallel with other environmental groups, Greenpeace has helped achieve results in these areas." Thus, although the International Fund For Animal Welfare must receive ultimate credit for the Common Market ban, Greenpeace would have it supporters believe that *its* action was of primary importance since IFAW is not mentioned here.

Greenpeace claims to have influenced the United Nations through "passage of three resolutions calling for an end to all nuclear weapons testing," and continues to demonstrate in protest of nuclear missile bases and test sites internationally. Greenpeace claims credit for "a substantial reduction in the quantity of titanium dioxide wastes dumped off the Eastern United States into the Atlantic Ocean, and an indefinite suspension of ocean incineration of hazardous wastes in the Gulf of Mexico."

All of this sounds extremely impressive, and in the absence of any other information on any of these subjects, most individuals whose consciousness had thus been raised would feel admiration for Greenpeace.

The accomplishments listed in the Greenpeace USA brochure are not necessarily to be credited to the American chapter exclusively, but more likely were exploits of the international organization. The name "Greenpeace" is legally controlled by the International Greenpeace Council, yet chapters which describe "their" activities apparently include those known through media coverage of the aggregate in order to take advantage of the fame earned by Greenpeace.

Greenpeace USA was formed in 1979 and claims supporters numbering 265,000. It has 501(c)(3) charitable organization status with the Internal Revenue Service, so its supporters' donations are tax deductible. However, a 1982 report of the Better Business Bureau found that Greenpeace USA did not meet its standards calling for "an active governing body meeting with reasonable frequency and an independent governing body whose compensated members constitute no more than 20% of the total voting membership. Greenpeace's board met twice (in 1981) and all eight of its board members were compensated."

Those who support Greenpeace and believe in its claims and goals, feel that it is a wholly admirable organization. These people will not be influenced at all by the negative opinions of the Better Business Bureau, or its Philanthropic Advisory Service.

Greenpeace has become a culture hero for the western world and is seen as part of a growth industry. The ordinary person seldom questions its involvement in any issue, unless that person feels personally threatened by specific Greenpeace action.

One example of such a threat to the ordinary person is the Greenpeace move to prevent elk hunting in the western United States and Canada. Some Greenpeace members have been said to spook hunters' pack animals, verbally abuse and (allegedly) threaten hunters and their wives and guides while in the field, and in general have attempted to ruin many hunting opportunities. Such behaviour has been featured in *Field & Stream* and other outdoor magazines, and occasionally is taken up by major news media.

Some native Americans hold Greenpeace in especially low regard: Alaskan Inuit have felt that their taking of bowhead and other whales is their own business and that outsiders such as Greenpeace activists should not try to interfere in this part of their lifestyle. Since Greenpeace wishes to take credit for ruining the market for sealskin in Europe, it should also expect Canadian and Greenland Inuit to look upon them with disfavour as sealskin has always been of vital importance to Inuit as trade material and as an independent source of income.

Interestingly enough, a poignantly ironic twist has come about because of the loss of the sealskin market; Greenland Thule Eskimo who used to derive a very significant portion of their income from the sale of harp sealskin, turned increasingly to hunting the narwhal. A *National Geographic* article on their lifestyle highlights the situation which led them to concentrate on finding adult male narwhals with long, perfect tusks. Such trophies brought $800 cash for the hunter and the money was used for those things of daily life which used to be derived from the sale of harp seal. In 1984, in response to this, the Common Market Parliament passed a ban on the importation of narwhal tusks.

Greenpeace supporters should realize that their many donated dollars may be having an effect quite beyond their comprehension or intent. "Saving" a favoured species (harp seal) resulted in economic hardship for the native hunters who then turned to another favoured species which was exploited for its salable tusks. Although the whole narwhal was used, large males were sought because of the economic advantage they would bring. This was absolute survival necessity, brought about by anti-sealing protest thousands of miles and eons of cultural realities away from the Arctic.

Greenpeace policy has long taken a position of protest against the seal hunt for reasons which it claims are humane and "moral." A major focus of the Greenpeace argument against the hunting of whitecoat and blueback pups has apparently come down to the fact that the culture which spawned Greenpeace is averse to the killing of infant creatures. Many people in the western world are easily swayed by any suggestion that it is not right to kill "babies." Thus, Greenpeace used Bridget Bardot in its film *The Rites of Spring*. Ms Bardot compared "fair" hunting with the harp seal pup hunt and avowed a significant difference: "You never kill the babies, never."

Patrick Moore, formerly head of Greenspeace Canada, has often made the point that it is wrong to kill "nursing marine infants on the breeding ground" as though this was a universal rule which had been violated by the Canadian seal hunt. His suggestion that the pups are still in the nursing stage when killed is not totally incorrect. Many are within a day or two of being abandoned by their mothers. Since the entire herd in any one place tends to whelp over a two-week period, there are apt to be seals on the ice which range between newborn and already weaned and abandoned by the time the sealers come on large ships or (in the Gulf of St. Lawrence) on longliners run by landsmen. Thus, the "nursery" is indeed entered by humans while some pups are still nursing. However, the men who come to harvest sealskin know that the most economical behaviour is that which concentrates on the older pups; they have a fully white coat and a heavier load of fat. The pelt is overall larger than that of pups which are still nursing regularly and is worth more.

Pups and their mothers are not terrified and brutally separated by force, as has been suggested by protest. Most adult female seals leave the ice when the men come. Those pups which are attended by persistent dams are left alone, out of deference to suggestions in the sealers' manual and the A 1 training film that they not be disturbed. Regulations state that it is not legal to hurt or kill an adult seal on the whelping ice.

Therefore, a mother seal is not harassed if she should be the rare individual which stays in defense of her pup. It is simply not worth the effort and the risk of being arrested and losing one's license for the season. Enforcement *is* a constant reality, both for landsmen and for commercial ships' crews so it is not reasonable to argue that much goes on which is unseen and unreported.

There is no indication that mating behaviour is disturbed or prevented by the taking of whitecoat or blueback seals. Copulation takes place in the water for harps, and on the ice for hooded seals, and no evidence has been presented that this is affected by the pup hunt.

There is no evidence that any seals abandon their unweaned pups due to the commotion around them of humans taking other pups by clubbing or other means. Therefore, no small nursing pups are affected by the hunt, which has passed them by.

Mother seals abandon their pups by the tenth day of life. They do *not* weep tears of grief (as has been claimed by the IFAW). There is no indication that nursing adult harp seals are particularly affected by the loss of a pup two or three days before it would have been abandoned naturally. Their behaviour has not been noticeably different from that of dams which had already abandoned their pups. Both come back to the whelping patch to rest before they come into estrus. Neither stays very long after the hunting activity is over. Mating takes precedence in their behaviour almost as soon as the pups are ten days of age.

The Greenpeace claim that it is "wrong to violate the sanctity of the marine nursery" is a cultural perspective, not a statement of actual harm done to marine mammals in the sense of affecting the welfare of the species. There is no evidence that cruelty, either physical or emotional, is a real result of the pup hunt. Greenpeace and IFAW films which allege to depict mother seals being hurt or killed in defense of their pups are not a reflection of seal hunt reality in the '80s.

The ban on the taking of adult seals, aggressive or not, on the whelping ice has been the result of scientific recognition that they are important to the herd from a conservation standpoint. Their reproductive potential is so great, compared to that of a new pup, that the ban on killing them is seen as an important part of promoting herd strength.

Greenpeace policy in the '80s has been to decry the taking of any seal under the age of one year. This apparently is a logical progression of their claims that it is wrong to take "defenseless" pups. Sexually immature seals, however, range from newborn to approximately age four. If it is "moral" to take only mature animals, then by this reasoning only animals age four or over should be hunted.

Due to colour patterns which develop by the age of two, however, it is impossible to differentiate the age of any harp seal with certainty until it is killed and its teeth examined for growth layers in the enamel.

Thus, there is no practical way to decide if a seal is to be taken if Greenpeace standards are applied. The view of marine biologists (and of the World Wildlife Fund) is that herd welfare is *not* endangered if the majority of seals taken are *under* the age of one year. Too many seals over the age of one year might be taken, if Greenpeace criteria for "moral hunting" were suddenly applied.

There is no rational reason, based on seal welfare, for the protest against the taking of immature seals. The argument that they should not be taken because they are "defenseless" is one based on those particular emotional standards of an urban society. There is a feeling in that culture that it is unfair to take an animal which cannot flee, or which is "unwary" or unaware of danger.

Harp seal pups do not seem to have much (if any) fear of man while they are still in the whitecoat stage. They do not often try to avoid people, but are easy to approach and to kill. This in itself is something which terribly upsets those who come to photograph the slaughter and protest against it.

Such concerned outside observers might ask themselves if they would feel any better if white pups made a great effort to escape being clubbed, or if they seemed to have a real fear and apprehension at hunt time. A behavioural specialist could point out that the pups are better off as unwary and rather stupid, dull creatures. They do not seem to feel any fear or stress compared to domestic stock or poultry about to be slaughtered. Their "defenseless" state, then, can also be viewed as an "oblivious" one.

Those who have seen their photographs with wet eyes, soft fluffy hair and bristly whiskers and termed them "highly sensitive" are mistaken. This writer was very surprised to find them attractive, yet so apparently dull-witted and relaxed that they displayed no flight response or other signs of fear.

If the harp seal pup hunt was actually a sport during which one attempted to take a feisty, wary trophy animal, then there would be no doubt that the taking of "defenseless" infant pups would be unprecedented and inexcusable, non-sportsmanlike. This is the cultural view of sport hunting and it does not apply to the taking of these seals off the eastern coast of Canada. There is no "sport" in clubbing a small seal and no sealer ever claimed that "sport" had anything to do with his being out there.

The taking of seals is a welcome change from winter do-nothing boredom for many of those men who participate in the hunt, but this is not a valid reason to criticize them. The entire cultural premise of the seal hunt is one of necessary subsistence activity. This has always been the case.

The only instance of seals being hunted for "sport" is seen when outsiders come with center-fire rifles to try their marksmanship (for adults when it is not breeding season) with a legal sport sealing license granted by Canada. These persons may take adult seals, the meat and hides of which must be turned over to a native guide who greatly appreciates such a regulation.

The annual pup hunt bears no more cultural resemblance to legal sport hunting in other climates than does subsistence purse-seine fishing compare to the weekend fisherman's quest for tarpon, muskelunge, or trout. There is thus no justification for criticizing it from such an inappropriate perspective, and this should be pointed out to protest supporters who feel justified in using "pup defenselessness" as another reason the hunt should be abolished. The argument can thus be dismissed as another example of classic ethnocentricism.

Greenpeace, the Humane Society of the United States, the Animal Protection Institute, the International Fund For Animal Welfare, the Seal Rescue Fund, the Fund for Animals, and all other such groups which have protested the hunt have let it be known that at least in part their position is to disapprove of sealing which takes "baby" animals. All have implied that this focus is "unfair," although they also object to instances of sport hunting of other species under other circumstances. If one personally hates to think of infant animals being killed for any reason, then it is natural for that person to object to the hunting of harp seal pups, especially if that hunt is described as cruel and if photographs of the infant "victims" are frequently a part of the argument.

The culture of the western world is so firmly pet-oriented and humane-treatment-oriented that every attractive species, wild or domestic, is regarded as a pet object in its infant stage. It would appear that the protest movement has fully realized this and has purposefully enhanced it with modern photography and skillful fictitious depictions of the harp seal. The image of the white pup being held "protectively" by a caring human (member or leader of a protest organization) is one of the most familiar shown in the course of protest. It is routinely contrasted with photography of "the bad guys" clubbing such pups for their evil, commercial reasons.

It is no wonder that the sealers of Norway, the Artic, and Canada are held in such low international regard. Protest groups know they shall continue to be supported by a kind, caring public if they oppose all sealing everywhere. Such a position is a perfect cultural fit, causes the supporters no inconvenience or worry, and reinforces their high opinion of the humane standards of their own society. It also effectively precludes a climate of cross-cultural tolerance and good will. Those protest supporters who are Christian have perhaps not yet perceived that such protest is contradictory to world fellowship and might wish to think about it as a problem.

The Greenpeace concern over the seal hunt has also been extended to the sealers themselves, for both humane and social reasons. The organization has claimed that Canadian fishermen would benefit more from receiving a government subsidy payment *not* to hunt seals than from the yearly risks which must be taken in order to profit from sealing.

Subsidy is not a new idea; versions of it have been offered by other protest organizations, all of which have claimed concern for the emotional and economic well-being of the sealers themselves. Greenpeace has claimed that such a subsidy should be readily acceptable to Newfoundlanders since many already accept wintertime unemployment insurance cheques as a routine way of surviving the stretch between fishing seasons. However, there are several reasons why the idea has not been well received by those it is meant to help; a common claim made by Greenpeace and other protestors is that sealing "degrades and dehumanizes" those who participate in it.

Given this denouncement of the culture by outsiders, it follows that if any form or source of subsidy payment was accepted in lieu of sealing, this would be admission on the part of all sealers that the practice was indeed dehumanizing, degrading, unsavory, undesirable, and unnecessary to survival, happiness and pride. The subsidy idea in any form has always been soundly rejected by Newfoundlanders and others who routinely hunt seals in eastern Canada.

Part of the reason for this has been a lack of confidence in the continuing nature of the subsidy, and part has been pure cultural pride in their tradition. Those who have always gone sealing fail to understand why anyone should reasonably expect them to give it up. The sealers absolutely reject the notion that the activity is degrading and harmful to themselves or their environment.

An anthropological perspective of this clash of cultural values has never been brought out by the press with any thoroughness. The ethical problems associated with this protest attempt at forced cultural change have not been addressed. The tendency of the cultural anthropologist would be to view the protest movement as entirely unethical due to this blatant disregard for human impact, and its explicit denial of the intrinsic worth of Atlantic culture.

Greenpeace cannot be excused, within this framework, by claims that it never meant to hurt those who hunt seals or whales for subsistence reasons, but just to obstruct those who profit commercially, "make useless trinkets," "sell trophies," or "luxury furs and leathers." There is a logical inconsistency to the argument, because for an Inuit or a Newfoundlander or a Quebec citizen, subsistence hunting means just that. This is not a sport for them. Meat to be eaten, fuel to be purchased, and bills to be paid are, to those people, equally important. The fact that the prey objects are not endangered is relevant. If they were, then it would be the duty of the government to protect them from harmful exploitation. But since they are demonstrably not endangered, it is certainly not the province of outsiders to object to their use.

Further, the type of lifestyle supported by this hunting activity is irrelevant to the moral argument that those people should be allowed to continue it. One need not live miserably in order to justify one's decision to hunt marine mammals in order to survive. It does not matter whether the hunters trade for clothes and medicine at a "post" in the Arctic, or buy anything available, necessary or not, with cash in a modern supermarket or shopping mall. If this is their cultural reality and their cultural choice, they should not be shamed or coaxed or coerced into doing otherwise by affluent outsiders with different value systems. Igloos, sod huts, kitty litter, colour televisions, triple storm windows and microwave ovens are not the business of Greenpeace administrators or foreign journalists who come to judge whether or not the "natives" should be hunting seals. Their needs are their own business, and have nothing to do with the biological impact of their hunting choices.

The Greenpeace proposal for "solving the problem" of the seal hunt has been to suggest that the Canadian government officially abolish it and, since fishermen receive unemployment insurance each winter, continuation of this in lieu of sealing would suffice just as well.

Dr. Patrick Moore, formerly head of Greenpeace Canada and more recently a member of the Board of Directors of Greenpeace International, has publically stated that the fishermen should not mind taking an additional two months of UIC payments since they are not too proud to accept them each winter, anyway. This sentiment has been echoed by the spokesperson for Greenpeace USA and so it may be assumed that the policy is a general Greenpeace one and not limited to any one chapter.

This attitude infuriates those fishermen who know that for many, unemployment cheques are the *only* way to survive the winter and are not taken through laziness or unwillingness to find employment alternatives. There are no employment alternatives in fishing villages in the winter. There is insufficient power and few other resources to allow for development of a factory system and there is a very wide dispersal of the population, spread thinly up and down the irregular coastline.

The fact that Greenpeace seems not to distinguish between unemployment cheques and "welfare" and that they seem to imply that fishermen would not mind an extension of such "welfare" in lieu of going back to work during seal time, gives Dr. Moore and the Greenpeace name a very high profile on the east coast. Most fishermen feel that Greenpeace representatives would be wise not to visit their shores again.

Unemployment cheques are computed on the basis of a percentage of each man's income throughout the year. If the fishing season was poor, the unemployment rate is low; in any event, the cheque is less than was his monthly income. Therefore, most men are eager to give up their cheques as soon as the seals come in. There is no practical alternative. In a good year more than twice as much can be made by sealing as by sitting home and drawing unemployment.

Greenpeace has suggested that each man and his family would be content with a dribble of fixed income while the seals grow in number and eat increasing amounts of cod and caplin. Canadian fishermen want to tell the world that the Greenpeace idea is based on a combination of ignorance and unprecedented audacity.

There has been no test of Canadian taxpayer willingness to force fishermen into acceptance of a subsidy in lieu of sealing. Chances are many Canadians would rather not pay them to sit around when they might be out at the ice and "off the dole." Greenpeace Canada may be wise enough not to try to force this to a test if the market for sealskin ever revives.

There is a basic distaste in western culture for any doctrine which gives the work ethic such a low priority. Fishermen have always been known as hard workers who take great personal risk and have realized little enough profit for their efforts. There has never been a serious suggestion that they give up fishing because outsiders have considered it a nasty profession, or because fishermen make much less money than do those who wholesale or retail the product.

Yet this argument has been made in support of abolishing the seal hunt. The profits, say the protest movement, go mainly to the fur traders and retailers. The fishermen who produce the raw material should get out of the system because they are "exploited." The same argument could easily be made about the economic plight of those who raise beef, pork, or chicken, but who would support it? Certainly not the consumers, and certainly not the farmers.

Those who are the consumers in the fur trade are a relative minority; therefore, the logical and cultural inconsistencies of this protest against the use of sealskin as a resource have not been forced into public attention.

Perhaps one reason for this has been a feeling that the wearing of fur is a luxury compared to the consumption of meat. This is only a matter of opinion held by the fortunate in the western world. If considered against the backdrop of third world famine conditions, it would be reduced to unimportance.

Greenpeace argues that fishermen should not kill seals because in so doing, they are being exploited. This exploitation is in the form of disproportionate profit when the industry as a whole is considered. Greenpeace also claims that the act of sealing is terribly degrading and dehumanizing for anyone.

Fishermen who go sealing vehemently disagree. At present the profit problem is seen by them to be the fault of the protest movement. The act of sealing is viewed by them as hard and dangerous work which is personally and culturally rewarding.

The Greenpeace perspective, then, is reduced to the level of ethnocentric myopia. It is a false argument made on behalf of people who do not appreciate these concerns of fishermen who do not wish to give up sealing.

On this part of the seal issue, an independent outside observer would have to conclude that Greenpeace arguments do not ring true and that proper attention was not given by that organization to the cultural and economic realities of eastern Canada and the Arctic in general.

Newfoundlanders are especially sensitive to another problem which they feel is directly related to the activities and campaigns of Greenpeace; the long years of Greenpeace concern for their "humanity" has led the international public to believe that sealing has, indeed, dehumanized the sealer. Newfoundlanders have been routinely termed brutes, thugs, and insensitive environmental vandals. There has been a move in England to portray them as monsters, on a par with the worst of the Nazi war criminals.

Although Greenpeace may not be able to take as much credit for this as the International Fund for Animal Welfare, which describes its view of the hunt in graphic and lurid detail in its mail campaigns, the blame for this horrendous image of the Newfoundlander may be a toss-up between the two organizations. Certainly the logical progression may be traced to Greenpeace. The group feared that sealers would be degraded and brutalized by the act of sealing. So, the public now believes that anyone who clubs "baby" seals is a sub-human monster.

Almost any social scientist would recognize this device of defamation of another social group as a neat ploy to ensure that one's own values are strengthened and upheld and that the other group's values and behaviours are seen as inhuman, unnatural, and wrong behaviour.

Of course, it would not be constructive to accuse the Greenpeace administration of consciously using this device to convince potential supporters to see sealing in the proper light. There is no way to prove such planning, and it doesn't matter anymore. The damage is done. Newfoundlanders are known the world-over as baby-seal-killing thugs.

What damage can this do? one might ask. Newfoundlanders don't have the money to travel that much and an occasional foreign newspaper article should not make much difference to them.

It made a great difference to their children in the mid '70s. A world-wide pen pal programme was instituted in the schools. All the were supposed to pick a pal from a list of foreign students of the same grade level and introduce themselves by writing about their family life.

Those Newfoundland children who admitted to having fathers who were fishermen received no answers. Their "pals" in Great Britain and Europe perhaps suddenly realized that they had been put in touch with undesirable people. Or, more likely, their parents and teachers suddenly realized it and found them more suitable correspondants. One Newfoundland girl, now twenty years old, said that she and her friends could not understand, at the time, what had happened. Her only respondent was a boy from Hong Kong who did not seem to care that she was a fisherman's daughter.

The entire international pen pal programme collapsed in Newfoundland. What had begun as a hope for introducing these children to their part in the world community ended in disillusionment and confusion. They never knew that it wasn't somehow their fault. There was no way to help them to feel any better about it at the time. When they grew older, they began to understand, but the hurt remained. Their parents say little, but know that the protest movement in general, and Greenpeace and the IFAW in particular, are responsible for the international view that Newfoundlanders are somehow less than human.

This unfortunate social consequence, however, has not caused any change in Greenpeace policy statements. The Greenpeace view is that all commercial sealing is non-subsistence sealing. Thus Greenpeace is against all sealing by anyone who is not an Inuit and is against even *their* participation in the hide market. The landsmen sealers as well as those who hire onto large ships to take whitecoats are seen by Greenpeace as unjustified in this quest which "supports only the international fur trade." The fact that protest has put an end to the major world market for seal products, thus doing great harm to Inuit as well as Newfoundland sealers, is seen as an unplanned and unfortunate side effect.

Greenpeace claims to have researched ways for "some of the hunters to earn their living WITHOUT killing seals." This probably sounds good to Greenpeace supporters, but the truth is that no seal hunters have been "helped" or "persuaded" by Greenpeace to find other ways of making a decent living in that time between the end of winter and the beginning of the ice-free fishing season.

A Greenpeace USA mailing in April and May of 1984 (Greenpeace Information on Harp and Hooded Seals) includes the following statement of purpose:

> FUTURE ACTIONS
> Greenpeace will not slacken its efforts to stop the seal slaughter, even though it has been greatly reduced. We will continue to educate consumers in the countries still buying products, and countries where sealers are seeking new markets, such as in the Far East. Greenpeace is also working to get the European Common Market to pass a special resolution that would extend the EEC ban on the import of seal pelts to include pelts of all seals under one year of age. This would have the effect of destroying what is left of the commercial harp seal slaughter, assuming that no new markets are found.

At present, the Common Market ban is on the skins of whitecoats (harps) and bluebacks (just-weaned hooded seals) only. The skins of beaters and older seals are not covered under this ban.

The Greenpeace organization apparently feels that the public has supported the "save the seals" effort because it was concentrated on "saving babies." The present move to keep the protest going is a logical one, since it is extended to "saving" all seals from commercial exploitation, whether they are "on the breeding grounds" or not. This may be in response to the realization that the IFAW is not about to give up on a good thing, now that the whitecoat ban has taken place and may well be extended indefinitely.

How would it look, after all, if Greenpeace did not extend its attention to all ages of seals, given the fact that the IFAW, a major competitor for funds, is continuing the protest in this manner? It is reasonable to expect that as protest groups win such major victories, they may feel they have to go far beyond their original goals in order to remain viable and keep up their credibility with their constituents. If the final battle is won, there is no more raison d'etre.

Supporter participation is said to be a vital part of the effort. In the spring of 1984, Greenpeace USA advised:

> WHAT CAN YOU DO TO HELP SAVE THE SEALS
> * COPY THIS FACT SHEET AND DISTRIBUTE IT TO YOUR FRIENDS AND CO-WORKERS. WRITE GREENPEACE AND ASK FOR PETITIONS AND POSTCARDS. DISTRIBUTE THEM AS MUCH AS YOU CAN.
> * WRITE YOUR NEAREST CANADIAN AND NORWEGIAN CONSULATES OR EMBASSIES. TELL THE CONSUL STAFF YOU SUPPORT AN END TO THE SLAUGHTER.
>
> Royal Norwegian Embassy
> 2720 34th Street N.W.
> Washington, D.C. 20008
>
> Embassy of Canada
> 1746 Massachusetts Avenue N.W.
> Washington, D.C. 20036

The offices of major Greenpeace chapters in the United States were then listed so that cards and petitions could be requested, and donation sent.

One Greenpeace claim is that all of its own protest is non-violent, although it may well be directly confrontational, as when volunteers dye seals or hinder whaling ships by zipping their zodiacs between harpooners and whales. This emphasis on non-violence is an effort to maintain a non-terrorist image in the face of criticism and to convince those potential supporters who abhor violence for any reason that the forces of good can prevail without causing real harm.

Therefore, when Greenpeace volunteers are arrested while spraying seals, or while spying on Russian whaling ships that allegedly supply meat (from non-endangered gray whales) to their mink farms in Siberia, or while blockading employees from entering a nuclear power plant, they are seen (and treated) by the press as "environmentalists" just trying to do good deeds and raise the public consciousness of large-scale evil.

In each case there is some emphasis on the personal discomfort or danger which Greenpeace volunteers undergo in order to accomplish their mission. Being arrested on camera is the frosting on this media cake. The organization will bail them out, and pay the fines and court costs. The public will compose letters of protest to those who arrested them, and decry the "seal protection regulations" or whatever under which they were charged.

Greenpeace is an international media star, supported by perhaps a million people who send money to their own national chapters in order to keep the good works going. International credibility is based on its "whole earth concern" message. "WE ARE ECOLOGISTS" is the key to success. A sufficient number of the public appreciates this approach to problem-solving on its behalf and keeps the organization solvent. For Greenpeace supporters, each individual issue's credibility is enhanced by association-through-protest with all the rest.

This high profile by Greenpeace did help in influencing Europeans to press for a Common Market ban on the import of seal pup products. Therefore, it has to be said that Greenpeace has had an effect on the Canadian harp seal issue. Greenpeace is second only to the International Fund For Animal Welfare in influencing world market conditions and world opinion. Its supporters continue to donate for the good of the planet.

When one considers the nature of Greenpeace action globally, it is seen to be pretty safe for the organization. Greenpeace has not risked the embarrassment of failure the way the IFAW did with the fish boycott. The Greenpeace way is to ask for public support of large general issues and ideals, which will in no way cause the membership to suffer personal economic hardship or real inconvenience. In general, Greenpeace avoids major, expensive campaigns such as the IFAW's world-wide boycott of Canadian fish. The Greenpeace way is relatively risk-free environmental protest, and this may continue to be the Greenpeace pattern of success. Greenpeace raises public awareness of "the bad boys" and says of itself "WE are environmentalists." A comparatively low-key attempt to create a boycott of Norwegian fish in order to "save whales" is the only known exception to this rule.

One-third-page ads in *Time* magazine ask the consumer not to buy fish products labelled PRODUCT OF NORWAY, in protest. This is not much of a sacrifice for a whale lover, and doesn't compare to the IFAW's requests that Americans and British give up a wide range of fish products because the companies which produce them may have purchased them from Canada, "a nation of seal killers."

It is not likely that Greenpeace will ask its public to boycott Japanese goods, even though Japan continues to support whaling by its citizens. That simply wouldn't work and the failure would be a tactical error and an insurmountable embarrassment for Greenpeace. Such lack of support for a Greenpeace project would undermine its aura of serious protest.

A much safer approach is the current one which highlights Australian ranchers and farmers who kill kangaroos or who support their heavy cull for population control in the grassland. The Australian "Save Joey" protest uses photographs of an infant kangaroo peering out of the dead mother's pouch, and the words describing the slaughter sound amazingly like those describing the seal hunt. No humane person would be apt to resist a plea for funds for the kangaroo project.

Compared to Japan, Norway and Australia are easy targets of protest, so long as the public is not asked to give up anything substantial in the way of goods.

Greenpeace claims to support genuine research which is beneficial to problem solving in troubled environments. Since this is the province of national governments, it is not known how Greenpeace research can be applied to actual remedial measures in any one case. National governments research and solve their own problems, and do not accept data from groups which appear to have a vested interest in protest projects. Such data would be assumed to have a bias, since the organization is in the business of convincing the public to agree with one set of data and to donate in support of one perspective.

The actual amount of unbiased expertise used by Greenpeace in its "research" of each issue is not known. That portion of the public which supports the organization has no way to judge this. Few are apt to inquire into the hard data of Greenpeace policy statements.

In general, Greenpeace appears to invite one's support through a load of "facts" which one is encouraged to take on faith. The potential member or donor examines that literature which is particularly appealing to him or her, and by donating, in effect appoints the organization to do the thinking, lobbying, and protesting.

One can thus do good in the world yet stay safely at home and at work without risking anything. Due to the great credibility of Greenpeace few realize the potential for any harmful impact on innocent humans or on favoured species, or question that their support results in good works.

Greenpeace USA, Incorporated "is a not-for-profit organization which is engaged in saving endangered marine mammals and their oceanic environment." This quote is taken from a summary of significant accounting policies published by the Greenpeace accountant's statement for 1983. Those contributions which were received included some which were intended by the donors to go to the benefit of specific animals or projects. In those cases, separate accounts were set up for each project. The list of Greenpeace USA expenditures for the year July 1, 1982 through June 30, 1983 is as follows:

Seal campaign	$ 940,061
Whale campaign	402,149
Dolphin campaign	341,844
Toxic waste campaign	158,823
Endangered species	138,889
The Greenpeace Examiner (newsletter)	136,737
Comprehensive Test Ban Treaty	99,145
Marine ecology	75,299
Rainbow Warrior	42,109
Merchandise (manufactured for supporters)	311,158
Grants to other Greenpeace organizations	1,551,210
Total Program Services	4,197,414

The above list of expenditures must reflect Greenpeace USA values and priorities, and is also a reflection of the specific areas of support which its donors apparently value. Total support and revenue for that time period exceeded expenses by $244,829, and not all money donated for specific projects was actually spent on them; some remained in those separate accounts:

Deferred Revenue:	
Seals	$25,879
Whales	16,351
Dolphins	8,344
Toxic waste	1,273
Zodiacs	7,202
Endangered species	1,609

From this, it would appear that although Greenpeace USA claims to feel greatest concern for endangered species, its expenditures for all campaigns would indicate that its followers value other projects more highly. Therefore, one might conclude that the organization has shaped its supporters' priorities through its own "educational" efforts in ways that differ from its ideal list of priorities.

Many people who feel a great concern for toxic waste pollution, the idea of a nuclear freeze, and the fate of marine environments and their endangered species, might well look at the above list of expenditures and conclude that Greenpeace USA had failed to stress the big issues enough to generate meaningful income for them. Instead, the big spending push was focused on seals, probably including harps, which are not endangered. (Since endangered species is a separate category, perhaps seals and whales do not include "endangered ones"?) Not all whales are endangered, yet whales are the second largest expense. More was spent to produce merchandise (labelled "Greenpeace") than was spent on the efforts for the test ban treaty, marine ecology, or endangered species.

Revenue from merchandise totalled $368,334. Accordingly, the Greenpeace choice to produce such items resulted in a profit of only $75,176. The purchase by supporters of $386,334 worth of special merchandise was apparently more attractive in general than was the choice of a donation for a specific project or animal, except in the case of whales or seals.

This conclusion, however, is based on the assumption that the expenditures for whales, seals and other projects reflect rather closely the incoming revenues in that year earmarked for them. Perhaps, in all fairness, this is not the case. Perhaps the money was left over from another year and was spent in 1982-83 in response to a need perceived by the administration. Money for specific projects not spent, however, totals $60,658.

When a Greenpeace supporter sends any donation in to Greenpeace USA in response to a direct mail request for funds, "Greenpeace USA is required to pay over, as grants, to Greenpeace International net receipts from direct mail solicitations after deducting overhead expenses" (from the Greenpeace USA Notes to Financial Statements 1983). Therefore, while specific requests that one's donations go to certain projects may be honoured, just as stated in the Greenpeace USA list of actual expenditures, a large amount of money for which there was no requested use (perhaps this includes grants worth $1,551,210) goes to other chapters and the mother organization. Some donors may well wonder where or by what level of Greenpeace their specific "seal" or "whale" or other specified funds are used. Greenpeace International claims to have knowledge of exactly what is spent on every project by every chapter.

It is interesting, therefore, that the Greenpeace International office claimed $0 was spent by Greenpeace USA on seals in 1983 and that Greenpeace International, the

> ...primary funder for the seal campaign, expended approximately $110,000 in 1983 and $71,000 in 1984 in direct protest of international sealing issues. These funds provide not only for the Canadian harp and hooded seal hunt protest, but also for:

1. The Pribilof Islands fur seal hunt protest
2. New Zealand (hooker) sea lion depletion
3. Hawaiian monk seal critical habitat and entanglement issues
4. The Norway harp and hooded seal hunt
5. The Mediterranean monk seal endangerment
6. many other species suffering from population decline due to over-hunting, toxic poisoning, entanglement, and habitat destruction.

This rather amazing statement came in a letter from Anne Dingwall, researcher for Greenpeace International to a staff member of the Canadian Department of Fisheries and Oceans. It was part of the tiresome and sporadically on-going war of words between Greenpeace and DFO which never is resolved to anyone's real satisfaction.

In this case, DFO referred to a copy of the Greenpeace USA Financial Statements, prepared for them by the firm of Laventhol & Horwath, Certified Public Accountants for the year ending June 30, 1983. The statements given in that report (parts of which are quoted above) are apparently entirely different from those given to Greenpeace International, unless Ms Dingwall was badly mistaken. She went on to say that the figure of $940,061 "Exceeds the total Greenpeace USA expenditures for ALL campaigns by more than double." She stated that the figure was entirely inaccurate, and offered to send DFO accurate financial information for Greenpeace USA if such was desired.

Some terrible snafu apparently resulted in this misunderstanding among Greenpeace USA, the administration of Greenpeace International, and the office of the Canadian Department of Fisheries and Oceans about money spent on seal protest. At the time of this writing, the puzzle has not been resolved but it would seem fair to ask why the international arm of Greenpeace had no idea that the United States chapter had a publically disclosed, certified financial statement which differed so radically from its own balance sheets. Greenpeace supporters might wish to ask International why it claimed $0 was budgeted by Greenpeace USA for seals in 1983 when the USA chapter claimed it spent nearly a million dollars during that year on their behalf.

Direct mail fund-raising literature sent to the public in 1985 from Greenpeace USA now states that:

> GREENPEACE'S direct expenditures during the last three years on its seal campaigns were over $400,000. This money paid for actions resulting in the Common Market ban. It paid for consumer education and our continued worldwide lobbying efforts on

behalf of the seals. It also paid for the airplane that was sent to the scene of the slaughter, and for legal expenses incurred when the plane was seized and its crew jailed by Canadian authorities.

At this time it is at least fair to say that anyone might be confused by the above varying financial statements which perhaps should all be made available for comparative study.

The point of this discussion is not to hint at fraud or to imply that Greenpeace as an organization has no control over its finances. There must be a reasonable explanation for discrepancies of this magnitude and perhaps the entire thing shall be solved with the cooperation of Revenue Canada, the Internal Revenue Service, and Greenpeace International itself. If they feel insecure donors and supporters might hold off large bequests until such questions are publically answered. In the interim, perhaps a note to Ms Dingwall at the following address would satisfy supporters' queries: Greenpeace Northwest, Good Shepherd Center, 4649 Sunnyside Avenue North, Seattle, Washington 98103, U.S.A. (telephone 206-632-4326).

Over the years since Greenpeace was first formed, it has accepted, invested and spent millions of dollars in the name of environmental protest. This is not in itself rational justification for criticism, regardless of one's own biases for or against the taking of wildlife as a resource, or one's stand on acid rain, nuclear proliferation, or any other issue.

Many fine conservation or religious or charitable organizations take in and spend considerable millions in the course of their work. Some of them even lose track of their funds from time to time and some make unwise financial decisions. Most spend a very large percentage of their income on direct advertising for more funds and for general public support. All realize that pledges seldom meet expectations, and that a very small fraction of the public contacted actually donates money in substantial amounts. Ethical behaviour or relative merit of projects is not related to the amount of money which any organization manages to raise through the efforts of its professional, on-contract fund raiser.

Many people who have objected to the impact of Greenpeace or other such groups, have tried to discredit them by pointing out how much has been spent on the "propaganda" campaigns of each, how much is spent on fund raising in general, and the enormous sums which are routinely handled. This does not justify damning such groups. The same points could be made with the tax returns and financial statements of any large charity. The actual goals and aims of the organization in any case are apt to receive less funds than go to administrative and advertising activities. Creative leaders have to supported, must have travel and other expenses, and need to hire staff, public relations firms and marketing experts. Otherwise, no significant work will be accomplished.

If one really objects to the impact of Greenpeace or any other such organization, the way to make one's point is not by spending a great deal of effort on publishing its tax returns, but by causing others to think in depth about it goals and the means it uses to achieve them. If these can be criticized because they demonstrate basic wilful dishonesty, disregard for human rights, towering ignorance, anti-intellectualism, crass interest in profit at others' expense, or any other human flaw, then this is the area which should be exposed and discussed.

The only way to have a meaningful impact on the business of questioning high profile organizations is to make it possible for the general public to think rationally about them, based on all the information which can be gathered. All possible arguments and angles must be examined and scrutinized from as fair and well informed a perspective as possible. The basis for the current credibility of the organization has to be outlined and reasons why it is not justified, if they exist, must be made clear. Any factors which favour the claims to worthiness of the group must be listed. Thus the examination can stand on its own merit without being justly accused of bias.

Such questioning of the credibility of large and highly regarded charities or not-for-profit groups will always come under fire, if enough people pay attention to it. If this provokes thought and discussion, the process has been successful. It is only when a group and its ideals are accepted without question because of a neat cultural fit that intellectual activity and constructive criticism cease altogether. Then the human rights of those people whom the organization has targeted become of no concern in the overall process of protest, and this is wrong.

Some animal rights organizations decry "specieism" as an unwholesome tradition of thought held by the general public. Specieism appears to mean a preference for the well being of humans compared to that of other animals. A person who practices specieism does not object to the consumption of meat, the use of animals as beasts of burden, or to race horses, show dogs, zoo creatures, research rats, or pets which must eat on the floor. One who believes in anti-specieism is a vegetarian who disapproves of the use of any creatures by humans in any way which would be considered restrictive or confining or consumptive.

Such an extreme philosophy exists in a very tiny minority at present, but the current trend toward non-consumptive use of all wildlife (advocated by Greenpeace and many others), is a step in the direction of anti-specieism. Those who consider their own life style and think "it can't happen here" might consider that only fifty years ago no one believed that outsiders would ever put an end to subsistence sealing, "The greatest hunt in the world." In 1982 the sealing industry suffered an undeserved death blow. There is little reason to think that sport hunting or farming have a safe future.

The Sea Shepherd Conservation Society

There are a number of organizations which claim to be composed of "environmentalists"; most of these have large memberships, large operating budgets, excellent credibility with the general public as well as their own supporting memberships, and even legislative lobby arms. The Sea Shepherd Conservation Society is not one of these tremendously popular organizations, although executive director Paul Watson claims that his group is an "environmentalist" one.

According to the *Encyclopedia of Associations*, the Sea Shepherd Conservation Society claims only 5000 members, with six state groups. It is based in Vancouver, British Columbia, Canada.

The organization was founded in 1977, allegedly "right after Paul Watson was kicked off the Greenpeace Foundation's Board of Directors in June." (Watson was one of the original leaders in the Greenpeace movement, and apparently his own philosophies about the nature of direct protest action did not coincide with those of Greenpeace which has always claimed to be non-violent in its tactics.)

This was brought out in a CBC radio interview on the programme *As It Happens*, which was aired March 14, 1978. The present information is taken in good faith from a typed transcript of that radio broadcast, made available by a company named Media Tapes and Transcripts, an Ottawa-based firm. Although such typed transcripts often contain misspellings, and occasionally there are admittedly inaudible portions, the company which produces them does a public service by making them available to government and business when required. They are a valuable research tool which must always be taken with a grain of salt due to the possibility of human error in transcription.

The occasion of this broadcast was a CBC interview which included Frank Moores, then-premier of Newfoundland; Brian Davies, creator of the IFAW; and Paul Watson, who was apparently in a "telling" mood of pique about the condition of the world's environment and the role of protest organizations internationally. Watson was interviewed about the role of Greenpeace versus the IFAW in the protest game. His comments included the following:

"Well, I think that of all the animals in the world, or any environmental problem in the world, the harp seal is the easiest issue to raise funds on." He claimed that it was easy to make a profit on seal protest and that all organizations which protested the seal hunt made money with the issue. He pointed out that "there are over a thousand animals on the endangered species list...and the harp seal isn't one of them."

> [CBC:] *Did anyone in Greenpeace ever express that aloud, that it was easy to make some hay and some money on the seal hunt, so, therefore let's get into that?*
>
> [Watson:] *Well, yes, a lot of people have done that. You see, the seal is very easy to exploit as an image. We have posters, we have buttons, we have shirts...all of which portray the head of the baby seal with tears coming out of its eyes. Baby seals are always crying because the salt tears keep their eyes from freezing. But they have this image of...they are baby animals, they are beautiful. And because of that, coupled with the horror of the sealer hitting them over the head with a club, it is an image which just goes right to the heart of animal lovers all over North America.*

Watson freely discussed the fact that Greenpeace was a younger organization than the IFAW, but that it was rapidly catching up in its fund raising, especially on seals. "Last year I had submitted a budget for sixty thousand dollars. We spent fifty-five thousand dollars. And I believe that we raised well over a hundred thousand dollars." That was in 1977 and Greenpeace has come a long way in fund raising since that time. When asked how much he thought all protest groups realized annually from their advertising on an anti-seal hunt position, Watson estimated that it was probably three to four million dollars.

All of this information might seem more appropriately included in the section on Greenpeace, but the point to be made here is that Paul Watson had early experience with that major environmental group, and knew the possibilities offered by hunt protest. He knew how the game was played, and acknowledged that the public would pay more attention to the harp seal than to a less attractive and more deserving endangered dolphin or sea turtle. He said that there were "so many groups trying to raise money to protect dolphins and protect whales" that it was pretty hard to make money in that area.

When asked if the funds for support of protest were "collected from individuals who feel badly," or from "corporate givers," Watson replied "No mainly they are from...a lot of school children, a lot of pensioners."

> [CBC:] *Your fear is, then, that it isn't just money that people can easily spend, that it is coming from people you think who would be better off keeping it?*
>
> [Watson:] Well, I think that a lot of the money is now being abused.
>
> [CBC:] *In addition to their salary, I assume that there is a lot of money to be used from the group for your personal living expenses, travelling, money raising?*
>
> [Watson:] Oh, certainly. People, in addition to getting a salary, Greenpeace people are flying around the world all the time, Australia, Japan, Hawaii, California, Norway, England. At any time there are a dozen people that are on their way to or from these countries.

Watson went on to say that the problem with protest organizations in general was that "the organization becomes more important than the issue."

Since that 1978 Paul Watson has received a good deal of media attention for his protests and intervention with the taking of marine mammals as a resource. His position statements for the media have always been that the taking of marine mammals is immoral, ecologically unsound, and in some cases illegal. Watson has used his vessels to actually ram other ships at sea, threaten sealers on the ice, and as a serious threat to sealing vessels large and small.

His first command was the *Sea Shepherd*, the old ship which Cleveland Amory's Fund For Animals had purchased in order to go to the Gulf of St. Lawrence and protest the whitecoat hunt. Amory subsequently claimed that he "gave" the ship to Watson, who used it to ram a "pirate" whaling vessel which was thereby badly disabled. Watson also intervened in the slaughter of dolphins by the Japanese, repeatedly harassed a number of whaling ships internationally, and also claimed to have interfered with the government-sponsored shooting cull of gray seals off the coast of Scotland.

In a number of these instances, his behaviour was deliberately and definitely destructive. It was meant to be a serious threat of real harm to his opposition. Further, Watson has never denied that he will behave in a violent and destructive manner if and when he sees fit.

This has been Watson's choice of action, and it is where he departs philosophically from Greenpeace, which has a policy of intervention without actual violence. Each animal rights supporter who observes these two modes has to decide for him or herself if one or either is moral behaviour in light of all that has been discussed here on the issue of the use of marine mammals as a legitimate resource.

In 1983 Watson made the media spotlight again by claiming that he would sail to Newfoundland and the Gulf and ram any vessel which was headed for the seal ice. By this time he had command of the *Sea Shepherd II*. (The first old boat had been intentionally sunk by him after his ramming of a Spanish whaler, and when it appeared that the boat would be taken over by the authorities, on his arrest.)

Although his past history may seem entirely unlawful and utterly lacking in good judgement and concern for other humans at sea, the media treated Watson well in early 1983, giving him exposure on NBC's *Today* programme and feature time on *Real People* where Sarah Purcell appeared ready to break into tears as she considered his bravery and selfless concern for seals and whales. *Real People* aired Watson's segment several times so it must be assumed that the programme received a large and positive response from its viewers.

The *Today* programme aired on February 23, 1983 and Watson appeared in a peculiar "quasi-military" looking outfit. He was interviewed by Bryant Gumbel whose prepared introductory script was, as usual, coloured with sympathy for the "seal saviour":

> ...the springtime slaughter of baby seals has particularly always outraged wildlife conservationists. In fact, protesters of the hunt risk their lives to save those seal pups. Here [a film clip supplied by Watson is shown] Paul Watson and his colleague bodily block the path of a sealing vessel trying to make its way through the ice to the hunting grounds, and Paul Watson joins us now to talk about his plans to disrupt this year's hunt.
>
> [Gumbel:] *How do you plan to stop them this year?*
>
> [Watson:] Well what we intend to do is obstruct the three sealing vessels and prevent their entry into the nursing floes and we are prepared to go as far as ramming those vessels if need be to prevent that massacre from taking place again this year.

Watson went on to say that he was a "policing force" but Gumbel cut him off, saying "by whose mandate?" for, of course, Watson is not a law enforcement representative in anyone's country. Watson's reply was that since there was no international police force to protect seals "on the high seas," then he had the right to do so since the hunt had been "condemned" by the United States and by "seventy percent" of Canadians. He thus declared himself an "enforcer" in an area in which a legal hunting activity was scheduled to take place.

Watson's statement justifying his "policing" action included claims that Canada "refuses to allow meetings, discussions or the right to observe the hunt. Then the only path that there is open is confrontation and it's through confrontation that we have managed to achieve what we have achieved...that there is now no market for the seal pelts, yet the Canadian government intends to carry on anyway."

It may be noted that neither Watson's nor any other organization had conducted any "confrontations" in the campaign which led to the European pelt ban. The ban was accomplished by protest advertising which convinced the European public to apply pressure to its EEC representatives. "Confrontations" in Watson's sense of the word, meaning violent physical encounters between parties, were not a part of that protest action. Few Americans, however, knew the reasons for the pelt ban going through and so had no way to judge the veracity of these statements.

Perhaps Watson had problems obtaining permission to "observe" the hunt; one can only guess that if he had applied for permission, after having been arrested previously for protest activities conducted on the ice, he almost certainly would have been turned down. Those many individuals who have routinely been granted permission to observe the hunt have included serious marine scientists, writers, members of European parliaments, representatives of respected conservation organizations and a few "protest" organizations which had not previously intentionally broken the law on the seal ice.

It is not known how many *Today* viewers felt that his "reasons" justified Watson's 1983 threats to attack at sea any sealing vessels which he might encounter. His appearance on *Real People* must have been very satisfying for him, however. There he was billed as an environmental saviour, a real hero who gladly risked his own life in order to save marine mammals everywhere from death by harpoon or other deadly human invention.

Watson supplied film clips of whales swimming and singing underwater, interspersed with alleged clips of the *Sierra* preparing to harpoon one, then alleged footage of the *Sierra* being rammed by the *Sea Shepherd*. Stirring music and emotional narration completed the scene. His plans to "save" harp seal pups by threatening to ram any sealing vessels headed for the ice were outlined and film of harp pups was shown to demonstrate how deserving they are of such action.

After Watson's 1983 debacle on the ice of the Gulf, *Real People*'s postscript to the affair noted that he had been sentenced to fifteen months in jail but was out on appeal. Ms Purcell's voice quavered with indignation and scorn as she noted that Paul Watson had been charged with violating "The Seal Protection Regulations."

It is not known whether Watson sought out this hero treatment on *Real People* or if the producers came to him. Rumour has it that a major film production company has purchased the rights to the Paul Watson life story. It should be a stirring docudrama.

There is not much point in discussing Watson's philosophical stand on "saving" marine mammals from hunting. His reasons for doing it are ostensibly to "save" a certain few from immediate death, without regard for those who prey upon them. Watson, apparently, does not work to prevent wildlife habitat loss or damage, or support unbiased biological research.

In considering the tactics of the Sea Shepherd Conservation Society, one should remember that its founder believed two things: any policy to "save" seals and whales was a sure-fire fund raiser, and that at least some of the public would concur with violent, even illegal and deadly confrontation as a tactic. Perhaps Watson felt that this would be a media-grabber in contrast to the non-violent actions of Greenpeace, which he left in 1977.

There is always the possibility that Paul Watson is a kind, very concerned and moral man who wants to save seals, dolphins and whales at any cost, and has decided that his way is the only one which can be truly effective. At least 5,000 other people apparently agree and perhaps more shall, in time, join the Sea Shepherd Conservation Society in order to further this behaviour. Those who do not chose to do so may regard him as a "phony conservationist," but this opinion is not apt to deter him in his career, which will probably continue as long as he is out of jail and able to attract financing for his activities.

By late February of 1983 Watson had sailed the *Sea Shepherd II* as far north as South Portland, Maine. He apparently contacted one or more local elementary schools and offered to give school children a lecture on his "environmental concerns and tactics" and a tour of the ship.

It is not known whether school administrators were aware that the *Sea Shepherd II* had been boarded by the United States Coast Guard who had confiscated a small number of long guns and a rather large amount (allegedly some 300 rounds) of ammunition for one of them. Watson had previously made statements to the effect that, aside from his water cannon and an electric fence on board the ship, he was virtually unarmed. According to Canadian intelligence personnel who shared data with the United States Coast Guard, this was not the case. He also allegedly had some problem with proving that he had a licensed navigator on board and was informed by the Coast Guard that he could not leave port until one was obtained.

Nevertheless, a number of classes from South Portland and Falmouth, Maine, apparently led by teachers who were impressed by Watson's mission, boarded the *Sea Shepherd II* and were given a tour and lecture. Subsequently these children were instructed by their teachers to send thank-you letters to Watson. Some were on plain paper with drawings of the children's impressions, and some were on paper which apparently had been prepared and lettered by an adult, probably the teacher. These latter all had a mimeographed drawing of a harp seal pup on the top with a rainbow which was crayoned in and the mimeographed phrase: "Please let me live" next to the drawing. Letters were dated March 1, 1983. One teacher wrote:

> Dear Mr. Watson and Sea Shepherd Crew,
> These letters and drawings were done by third grade children from Falmouth, Maine. These children are eight and nine years of age, and are very interested in the plight of animals in the world. We are going to have a cookie sale to raise $15.00 in order to become supporting members of your society. I hope to make this a yearly project. I'm enclosing a stamped addressed envelope. If anyone on your crew has time, we would love to hear about your progress. Thank you for letting us visit!

Excerpts from a number of the children's letters are an indication of the impressions which their visit made upon them. Many statements are similar, and so some ideas which appeared more than once are given here as examples of points which apparently stuck with the listeners:

> Real People Wear Fake Furs!

> Dear Sea Shepherd Crew, I like harp seals and whales alot! I'm glad that someone thought of saving them or they would be extinct! I think it's a good idea to spray paint them green! The harp seal is my favorite sea animal. What are you going to think of next? Your friend,.... [signed by a little girl]

> We are glad to hear that some people are fighting the baby harp seal killings. Our teacher has been fighting it for 10 years. I've never known about the harp seal killings until this year. They are horrible! I'm glad someone came to their rescue. You are very kind to try to stop the Norwegian hunters. Good luck! Sincerely, a Concerned Citizen.... [signed by a little boy]

Thank you for saving the seal pups this year. I realy like you because it's nice to save animals that why i like your crewe. [A large blue whale is depicted, smiling, and a caption-statement coming from its mouth states "You saved me."]

Dear Captain Watson,
   I like what you are doing to stop the killing of the seal pups. When I get older, I'd like to work on a ship like yours. I like the idea of the Sea Shepherd sinking the hunter boats. I wish you luck on your next trip. Sincerely,..." [signed by a little boy]

Dear Captain Watson,
   I am very happy to see that someone cares about the harp seal. My class is very concerned about the harp seal. Keep on ramming those boats! I say you are brave enough to ram those boats and risk your life. Love,..." [signed by a little boy in grade five]

Another little boy wanted to know if the sealers whose boat would be sunk by the *Sea Shepherd* would take their clubs with them as they swam to shore, or if they would just jump in without them. This was a recurring interest. Apparently Watson spoke of the sealers as Norwegian and as "poachers"; several children said they did not like poachers anyway. More than one mentioned the ships of "the Norwegians," and Canadians were not specifically mentioned as seal hunters.

More than one child voiced the opinion that seals should not be killed just because they were in the way of oil exploration. It is not known what the information was which stimulated these comments from third and fifth graders.

One fifth grade child said, "I admire you ramming the Norwegian hunter's boats. Saving the seals is wonderful. If [you] sink a ship and someone drowned would you be convicted for murder? Why don't they move the seals to another cold place? Just because there is oil that's no reason to kill. I hope you agree. Sincerely,..." (signed by a little girl).

Some obvious concerns which a responsible adult might have about exposing small children to the Paul Watson perspective include the following: This person, who obviously espouses illegal activity and who has stated that he intends to break the law and to ram ships at sea, is presented to small children as a hero. The justification for this is that he means to "save animals" even though this involves a risk to human beings who are depicted by him as "bad." Such a perspective, encouraged by school teachers, cannot be considered responsible or commendable.

Apparently there was no counter perspective presented to those children before they wrote their thank-you letters. Perhaps no such perspective was ever presented to them in school since at least two teachers were mentioned by children as being entirely on the side of seal hunt protest. Thus, the protest movement was apparently presented as the only appropriate solution to what was perceived as a problem for wildlife. Alternative ideas about the sealing or whaling issues were unlikely to have been used in the classroom.

The children of South Portland and Falmouth, Maine may well grow up believing that violent confrontation is an apropriate way to deal with problems in the environment or elsewhere, as long as the problem-solver thinks he/she is right. The law enforcement perspective and the wildlife management perspective were apparently not included in the information which those children received at that stage of their education.

Children were obviously told that Watson feared the harp seals would become extinct through hunting. A number of them mentioned that he was "good" to try to prevent this by ramming the boats of the Norwegians.

Most children who enjoy violent programmes on television believe that no serious harm comes from the action; car crashes are exciting and funny to watch, and those that fly over banks into the river result in wet people, not in injury and death.

The children who wrote to Watson in support of his ship-ramming plans appeared to believe that the men on the rammed ships would merely "swim to shore," with or without their seal-killing clubs. Only one child mentioned the possibility that someone might drown, and then worried that Watson might be convicted of murder. Since Watson was the hero figure, many worried about *his* personal safety, however. Apparently none of the children were aware that death from exposure would be an almost inevitable and immediate result of human immersion in the icy water.

The above letters to Watson were found after the RCMP took over the ship, which he had abandoned to his quite frightened and apprehensive crew. Watson tried to escape arrest by walking on the ice to Cape Breton, where he was apprehended. The Maine children's letters remain on file with Canadian authorities. They were examined by this writer and felt to be a sad but authentic commentary on the teaching of social responsibility in the U.S.

In December 1982 a story had appeared in the San Francisco *Sun* to the effect that Oxfam, an ostensibly responsible agency concerned with third world famine conditions, had arranged with Watson to deliver farm tools and text books to Nicaragua and to the island of Grenada in early 1983, before heading up the east coast of North America towards Maine and Canada.

Canadian intelligence also alleged that there was some evidence that Watson planned to sell the *Sea Shepherd II* in Grenada after the 1983 seal escapades in Canadian waters. It is not known whether the connections Watson had in Nicaragua or in Grenada (later to be invaded by the United States) were purely benign and charitable, or if they also included political and military functions. A two-hundred-foot-long ship might reasonably be expected to have more potential as a military vessel than as a boon to agriculture and education in such poor nations. However, no proof of ill intention on the part of the Sea Shepherd Conservation Society in this matter has ever been offered to this writer.

After he left the port of South Portland, Maine Watson sailed the *Sea Shepherd II* to St. John's, Newfoundland, which has a small harbour with a very narrow entrance. No violent action against sealing vessels was taken by him inside or outside the harbour, although he waited outside for some time, allegedly vowing to ram any ships on their way to the hunt.

While he stayed offshore in sight of Signal Hill, which overlooks the harbour, Newfoundlanders by the hundreds drove up to the lookout to get a glimpse of the famous villain. On March 16, 1983 the St. John's *Evening Telegram* reviewed his past record of "action at sea" with the following accomplishments:

> SEA SHEPHERD LIES IN WAIT OFF NARROWS
> Now that Paul Watson's offer to pay sealers not to kill seals has been rejected, he and the crew aboard the *Sea Shepherd*, which is about five miles from the Narrows, are waiting for the sealers to leave port.
> He hasn't said exactly what he'll do, but he has threatened to try and ram the sealing vessels with the *Sea Shepherd*, as he did off Spain a few years ago causing $750,000 damage to a pirate whaler.
> The *Sea Shepherd*...also scuttled one of its vessels in 1980 in Lisbon, and eco-guerillas from the same organization sank two other vessels in the Canary Islands.
> Sources say Watson just missed hitting the explosives hold when he hit the pirate whaler, but he said the blow wasn't struck near the magazine, adding that the pirate whaler, the *Sierra*, wasn't carrying much explosives.

The article also noted that the act of ramming a vessel at sea is "defined as piracy in the Criminal Code of Canada."

Watson allegedly claimed that his proposed action was justified because sealers had turned down his latest offer of money in lieu of going sealing. The rejection, he said, "leaves us no other choice" but to continue lying in wait for ships on their way to the hunt.

Newfoundlanders are traditionally a people of the sea. They found Watson's threats and his presence both strange and somehow unreal, something almost so bizarre that it must not be taken seriously, and yet, because it was so unprecedented that any person could actually threaten to ram a ship, this was a frightening reality.

Reactions varied from demands that the government arrest him and his crew immediately, to hopes that he would just go away and bother someone else. There was, underneath, a cold fear that this apparent madman would use his ship to kill, and that nothing would be done until it was too late.

Since the government appeared to be doing nothing but wait for Watson to make his next move some people under the strain of suspense sought their own solutions. The plan to bomb the *Sea Shepherd* with a helicopter load of fresh chicken manure was the idea of Howie Hamilton, an accountant from Gander and president of Newfoundland's Fisheries Management Association. Hamilton's idea caught on, perhaps because of a need for comic relief and real retaliation. He said he was willing to put up $6,000 for the raid, but was unable to find a pilot willing to risk his license in such an illegal activity. The government has a law against "dumping at sea" except by special permit. Although Mr. Hamilton's representative to Parliament liked the idea and suggested that the provincial government provide a water bomber with which to carry it out, the raid did not go ahead.

A March 23 headline over a piece (by Michael Harris) in the Toronto *Globe and Mail* read: "Anti-sealers end siege, 'bomber' claims credit." The piece quoted Howie Hamilton as saying "They were committed to ramming our ships and I was committed to dumping a ton of chicken manure on them. It's as simple as that." Mr. Hamilton allegedly applied for the dumping permit, "but while he waited for an answer from federal authorities the *Sea Shepherd* sailed away." Hamilton attributed the departure to the fact that he had had a radio communication with Watson about the matter: "We flew over for a dry run on Sunday and then I got on the ship-to-shore radio and told them what was coming down. And I told them it would be very fresh — so the odor would linger."

On the subject of the apparent government inaction, Mr. Hamilton was quoted as saying "I just got fed up with politicians playing tiddlywinks with peoples' lives." His reaction with his own brand of direct confrontation was in frustrated response to what he perceived as inadequate law enforcement on the part of his government.

The fact that Watson did leave, perhaps not knowing whether the threat was a bluff, is interesting. He may have been ready to move on anyway. Or, the "bombing raid" may have been perceived by him as potentially disastrous press in which he would have been made to look ridiculous. The hero image would surely have been stained by these local "Newfie-counter-heroes" who refused to 1) take him seriously, and 2) be intimidated enough to sit without acting. Newfoundland won that one by appearing to back the *Sea Shepherd*'s captain off to safer waters, out of range of Howie's big bird.

The philosophers of the world may ponder whether Hamilton's proposed action was on a different moral plane than that of Watson. The possibility for human injury existed in both, but it was probably much less for the raid than for a ramming. Justification is another issue.

Watson had ended his siege of St. John's on March 23 and subsequently moved into the Gulf to begin searching in heavy ice for sealing vessels so that he could complete his "mission."

On March 26 Captain Morrissey Johnson of St. John's, Newfoundland moved his sealing ship, the *Clayton M. Johnson II* out of Catalina, which is farther north up the eastern coast of Newfoundland, and headed for the seals off the Front. He was perhaps by then assured that there would not be a direct confrontation at sea, although it is not known if he waited for that reason or because the ice, which had been unusually heavy that year, would not allow passage before that time. No one ever accused Morrissey Johnson of avoiding any anti-sealer.

By March 26 the *Sea Shepherd* was "under arrest for coming too close to smaller seal herds near Les Isles-de-la Magdeleine." The ship had been ordered to return to port, but Watson never acknowledged receiving that order. In that way he could plead innocent to a charge of failing to comply.

Watson claimed that an effort was made by the RCMP to lower an officer on board the ship from a helicopter, but that he (Watson) had told the RCMP that he would scuttle his ship if such an attempt were made. He claimed in a radio broadcast that "they're threatening us with some pretty heavy armaments."

Perhaps he was referring to the presence of Tracker aircraft which he claimed "buzzed" him "a half dozen times" and which were loaded with "anti-shipping air-to-surface missiles." According to the *Chronicle*

Herald, "Halifax Defense Department officials said the Trackers were not armed. What they're seeing is photographic equipment carried in a pod under the plane."

The captain of the *Sea Shepherd* abandoned it prior to its seizure by police at noon on March 27, 1983. Arrested were the remaining seventeen members of his crew, as well as Watson and three others who had fled across the ice to Cape Breton, Nova Scotia.

An account in the March 28, 1983 *Chronicle Herald* was typical of many versions of the event: the ice breaker *John A. MacDonald* had been brought in to effect the arrest, and a second ice breaker, the *Sir William Alexander*, assisted. Smoke bombs were lobbed onto the foredeck for a diversion and a gangplank was dropped from the ice breaker to break down the barbed wire electric fence which was strewn along the deck edges of the *Sea Shepherd*.

The St. John's *Evening Telegram* reported: "Arrest went smoothly says RCMP inspector," and this account stated that the boarding would have taken place while Watson was outside the harbour of St. John's but ice conditions prevented it. The boarding was effected with an RCMP Emergency Response Team, and the crew did not resist except for "six or eight persons who barricaded themselves behind a steel bulkhead door to the engine room where they were in the process of doing damage to the engine." According to an RCMP spokesman, the force was prepared for that and cut through the door with a torch.

The *Journal-Pioneer*, of Georgetown, Prince Edward Island, reported that the RCMP found that the "*Sea Shepherd* had been disabled by its crew before they were arrested," and that the disabled vessel could not be maneuvered properly, nearly ramming a ferry as it was brought into port. "The *Sir William Alexander* eventually brought the vessel alongside the dock and tied up beside her."

Another report out of Georgetown began with the headline, "Protest ship was strewn with garbage." When it was boarded, according to Fisheries Department spokesman Raoul Boucher, it was found to be in great need of cleaning and "It certainly doesn't look like a boat operated by an environmentalist." Other reports claimed that the bilges were full and that Watson had ordered no bathing, as water was in short supply. There was allegedly no heat on board, and little food.

Those who had reportedly paid Watson $1,000 each in order to enjoy being a part of the historic adventure may well have regretted their part in it all by the time their jail stay in Pearce, Quebec was over. They have not commented on this in the press, however.

The "Shepherds" were charged with conspiracy to commit mischief and conspiracy to commit extortion. "Justice Jocelyn Cote ordered the 16 men and five women held until March 31 for bail hearings" (from the *Globe and Mail*).

The crew were still in jail on April 6 and the *Daily News* headline revealed a startling truth: "Protest Group Can't Raise Enough Bail." Only two of the crew had been able to come up with the money, and the other seventeen remained in jail. Watson's wife, Starlet Lum, reportedly said, "We've spent everything we have on the *Sea Shepherd*."

As of April 6, 1983 only $33,500 of the $53,500 total bail had been raised to free the protesters. Ms Lum reportedly was "disappointed that other anti-seal-hunting groups had not contributed."

Watson was later charged additionally with piloting a ship in a dangerous manner, intimidation of sealers, mischief leading to endangering the lives of some sealers, and being unlawfully within a half mile of the seal hunt, a violation of the Seal Protection Regulations. Hearings were to drag on through the summer. Bail was finally paid, most of the charges were dropped, and the "Shepherds" were released.

The seal hunt went on in the Gulf of St. Lawrence, after having been curtailed for only one day due to Watson's disturbance. Watson claimed, however, that "72,500 seals had been saved" due to his action in the spring of 1983. He had arrived at the figure in the following way: according to the *Evening Telegram*, June 20, 1983, Watson said 3,700 seals were killed off Labrador and fewer than 9,000 were taken in the Gulf, for a combined total of 72,500 below the quota set for the year's hunt. He took credit for the reduced kill, stating that "sealers were intimidated by reports of terror tactics of the *Sea Shepherd*. 'Our campaign was primarily a bluff,' he said, assuring reporters he would never ram a sealing vessel." After his past performances, few people in Newfoundland took him seriously on that.

Watson was later tried and convicted in Quebec and spent Christmas of 1983 in jail, being released on December 29, 1983. He was granted leave to appeal his fifteen-month sentence for "interfering in the annual seal hunt in the Gulf of St. Lawrence last spring." He and his engineer, Paul Pezwick of Massachusetts, had been found guilty on December 20 of various charges relating to their adventure. Conditions of their release included being forbidden to discuss the case with any journalists and going anywhere near the Gulf of St. Lawrence between February 1 and May 1, 1984.

Watson had been fined $5,000 and sentenced to fifteen months in jail after his convictions regarding conspiracy to obstruct the hunt and causing mischief which engandered lives. He is currently out of jail on appeal. In the opinion of some, he has violated conditions of his release by discussing the case on television and with journalists.

During the week after Watson and his crew were first arrested and jailed in Quebec, the Toronto *Star* reported that a militant animal rights group admitted causing vandalism in the city and elsewhere, in protest against the arrest and incarceration. "Animal Rights Extremist Group Slashes Tires, Spray-Paints Walls" headlined an article which outlined a number of acts of vandalism which were explained

> In a letter delivered to the *Star* yesterday...The Animal Liberation Front admitted causing several thousand dollars in damage Monday night at the federal Fisheries and Oceans offices, two meat plants, and the Scarborough Animal Control. Metro police are investigating the vandalism, the latest in a wave of attacks by the British-based animal rights group.... Tires were slashed on a truck outside the A. Stork & Sons Ltd., a Queen St. W. poultry firm, and animal rights slogans were spray-painted on the walls. "These guys are screwballs. I don't know why they do this" (said a company spokesman). The slogans, "Fisheries Murder Seals" and "Set Paul Watson Free" were painted on the walls of the Fisheries and Oceans offices at Keele St.

Other businesses and animal research laboratories, et cetera, were also struck by this group and no arrests had been made. A caller told the Canadian Press in Ottawa that the incidents were "part of a continuing campaign of direct action against exploiters of animals."

It is not known if the Animal Liberation Front donated any dollars toward Watson's bail, or if there had ever been any connection or cooperation between the two groups. It may be reasonable to speculate that the ALF cared not a fig about Watson, but merely used the fact of his arrest for its own publicity purposes. There is no evidence for or against this theory, except that the other points which they wished to be publicized were not timely, in the sense that the seal hunt was, but were "old issues" with the ALF.

The *Sea Shepherd* remained in Canadian custody. Authorities claimed that it had definitely been maimed and although the "intent" of that damage has not been outlined in the press, the arresting force allegedly could hear the sounds of metal being damaged as they torched their way into the locked engine room where a small number of men were holding out. The crew denied doing any damage to the ship. By December of 1984 the ship was still in custody and final disposition plans had not been revealed.

Sealers remembered that Watson had offered them huge sums of money if they would stay onshore and not go sealing. They wondered where that money was supposed to have come from since Watson and his society had not been able to post their own bail. Watson had claimed that he would have been able to raise the money by holding a benefit concert of some sort. Skeptics suggest that if the money had actually been deposited in a bank before the offer was made some men might have been willing to gamble on it for one year, given the terrible ice conditions. Others, when asked, said they would never be bought, especially not by such a "phony" as Watson.

Whether or not this ex-Greenpeace, "eco-guerilla" is still speaking to classes of rapt elementary school children, or plans to appear on more home entertainment television programmes is questionable. He now has no ship, and with his past record of purposefully sinking or deliberately damaging other vessels, including his "own," it is not likely that he will obtain a loan to carry on with another vessel, or ever obtain insurance to sail again. The *Sea Shepherd* will be out to pasture for some time to come, unless there is a significant infusion of funds from his faithful flock to permit him to carry on his "humane" work. His fifteen-month jail sentence is still under appeal.

Anyone wishing to contact the Sea Shepherd Conservation Society may do so at: P.O. Box 48446, Vancouver, British Columbia V7X 1A2, Canada (telephone 604-688-7325).

The International Fund For Animal Welfare

The International Fund For Animal Welfare is based in Yarmouth Port, Massachusetts and boasts some 500,000 members. *The Encyclopedia of Associations* lists its purpose as: "To protect endangered species; to prevent cruelty to animals; to mitigate animal suffering; to promote cooperation among organizations having these same purposes; to solicit the support of all interested persons and corporations; to develop the interest of and to instruct the public in the work and purposes of the IFAW."

The IFAW publishes six newsletters per year for its members. These describe the projects and on-going interests of the organization and encourage continued support by the membership. Each one also contains the philosophy and world view of this segment of the animal rights movement and, like most "humane" publications, "exposes" the cruel behaviour of other cultures in their treatment of those animals which are favoured or pitied by the IFAW.

Although the statement of organizational purposes in *The*

Encyclopedia of Associations does not mention marine mammals, the IFAW has in fact been the principal force against the Canadian seal hunt. This is the only organization which has had a significant adverse effect on the market for sealskin and on the decision-making processes of those who are in the business of harvesting seals.

The IFAW as a non-profit organization currently (1984) has 501(c)(3) status in the United States which qualifies it for reduced mailing costs to its constituency. In the fall of 1983 and again in February 1984, a letter-sized, four-page information packet was mailed to the American public for a cost to the organization of five cents each. Reportedly, seven and a half million such packets on the IFAW fish boycott scheme were sent to Americans, each time at a savings to the IFAW of $1,050,000 over regular postage costs. At the five-cent rate, these separate mailings each cost the IFAW only $350,000 in postage.

Recipients were asked to be generous and to send the IFAW at least a $25 donation in order that more mailings could be sent, and that IFAW volunteers might go onto the ice again at seal time to save baby seals from the killers. A card enclosed with one boycott mailing contained on the left side a photograph of a whitecoat seal and on the right a donation form; contribution check-offs ranged from $10 through $1000 and "other" amounts, and it was stated that the contribution was tax deductible.

The Internal Revenue Service had taken the position that the IFAW was not eligible for tax-exempt status due to what it considered "substantial lobbying expenses," and it advised the organization that such status would be revoked on or about September 1982. Donations to IFAW may not have been considered tax deductible in 1983, but by the next year IFAW was once again on the IRS list of 501(c)(3) organizations.

In 1982 the Better Business Bureau found that the IFAW did not meet its standards calling for adequate controls over contributions because foreign donation records were not verifiable by the auditors who had been hired to compile financial statements. Also, IFAW did not meet the BBB requirement that soliciting organizations have an independent governing board in which no more than twenty percent of the total voting membership received compensation. The IFAW response was that it did not see any need to change its structure and it would continue to operate as it always had.

The Better Business Bureau concluded that donors should decide on the significance of these things themselves and should keep in mind that "an organization's practices may change at any time without notice.

This report is published solely to assist donors in exercising their own judgement."

At least one half million people in the western world now belong to the IFAW. It is probably fair to say that the negative opinions of the Internal Revenue Service and of the Better Business Bureau have had no significant effect on general public opinion of IFAW business affairs. Brian Davies himself has estimated that income from international fund-raising is typically in excess of six million dollars annually.

It would appear that, to the general public, the International Fund For Animal Welfare is an organization of vast humane concern. Few if any of its paying members would ever abandon it because of its "questionable" financial behaviour. They belong because the message of Brian Davies is completely credible to them and because it fits so well with their view of the way animals should be regarded by "good" people.

Since Brian Davies organized the International Fund for Animal Welfare in 1969 millions of people have come to believe that he would do anything to save seals from slaughter. Few know that, originally, he was eager to compromise, he wanted a trade-off. His plan was to have the Canadian government close down the St. Lawrence Gulf hunt and substitute for it a booming tourist business. When he felt that this might be a possibility, he spoke very convincingly to those members of the Canadian Parliament who had the power to introduce such legislation, and to provide for him the opportunity to become a rich man ferrying Americans out to the ice by helicopter each spring to see seals in their natural habitat. At the time he would have been willing to let Newfoundlanders hunt seals as usual.

During his testimony before the Standing Committee on Fisheries and Forestry, Davies never went so far as to say that he had purposely misrepresented the nature of the seal hunt to the public, or that he personally could ever approve of sealing conducted by anyone. Yet, in order to achieve his goal of creating a sanctuary-tourist attraction around the Magdalen Islands, he did make the following statements while under oath (spring 1969):

> In my opinion, sir, if you are going to hunt baby seals, the club is the least cruel way of doing it.
> Sir, I am satisfied that the sealing industry has improved its hunting methods in the Gulf — I think the hunt is less cruel now than it was.
> Yes, I think tourism is something that could be developed to a greater extent than it has been in the Maritimes using federal

> money...the killing of baby seals is so damned unnecessary, and it could be replaced.
>
> We do have an enormous population to the south of us, along this Atlantic seaboard...these people spend money just like water. I am convinced that a lot of people would pay the money to go out and look at seals. They like it. We can get people out to look at the seals in probably four or five hours from New York. It is just a unique opportunity.

Davies wished to make the point that the hunt in the Gulf could be stopped without causing economic harm to the inhabitants, through the substitution of spring tourism. When comparing this situation with that in Newfoundland, he stated that:

> I can really sympathize with the Newfoundlanders, because there is a tradition rooted in the seal hunt; their folklore has many heroes who went out to the hunt and died in a variety of brave circumstances and I am sorry that we find ourselves in a postion where we appear to be challenging this tradition...again, the hunt is merely dollars and cents. Nobody goes out to kill seals, I am sure, because they want to kill seals....
>
> Let me say that those of us who work for...and support this particular cause would like to see all harp seals spared wherever they may be because of a philosophical point of view that we have.... I think that I am close to public opinion.... I would have to say in all honesty that by and large world wide public opinion would — I do not like to use the word 'accept' but I just cannot think of a better one right now — accept a cessation of all seal hunting in gulf areas two and three but would not expect that Newfoundland landsmen who not only sell the pelt of the skin but very often eat the flesh as well, I do not think that public opinion would expect that these people not hunt seals in District No. 1 in the gulf area.

These statements indicate that Davies himself felt that the hunting of harp seals by those who need the products and whose tradition it is to go sealing was an acceptable activity. In 1969 sealing was acceptable to him if it *was* economically necessary and if it was being done in international waters where other nations could take the resource if the Newfoundlanders did not do so. At that time, he looked upon such resource use as justified, if not philosophically pleasing: "...I do not believe there is the slightest chance of creating a sanctuary for these animals on the Front and it is not a course that I would advocate anyone should spend money on in pursuing because I feel it would be money wasted."

When asked why he had decided to put so much effort into stopping the seal hunt in the Gulf, Davies said that the Artek film was instrumental in attracting his interest in the sealing issue. He knew that sealing was grossly misrepresented in the Artek film but he was also well aware of the amount and kind of response this would bring from the public. His own films on the seal hunt were used to shape public opinion and were considered by some in the Parliament to be a purposeful misrepresentation of normal sealing methods.

When asked the question, if the Gulf became a sanctuary, would it follow that public opinion would change so that a market for sealskin in the United States and Europe would recover, Davies replied

> I would guess that if the controversy died down, it would only make common sense to suppose that the market for seal skins would start to recover. Yes, it would only be common sense.

It may be noted that since that time he has ensured that the controversy would not die down by continual application of his anti-sealing message to an international public. The secret to belief, apparently, is a constant repetition of *any* information. Credibility grows out of repetition in the absence of other information on the subject, and this condition has been well met since 1969. The anti-sealing message has been echoed and enlarged upon by Greenpeace, The Fund For Animals, The Humane Society of the United States, the Animal Protection Institute, The Sea Shepherd Conservation Society, The Seal Rescue Fund, and the Society for Animal Rights, to name a few. Friends of Animals has used Davies' and the others' credibility to enhance its own campaign against sealing in the Pribilof Islands.

Brian Davies has focused on cruelty to seals. This is a "never fail" appeal in a society which has been pet-oriented for generations. IFAW literature has fostered the basic idea that the good person is against the hurting, hunting, and any other killing which men may do of any wild animal for any reason. Davies' faith in this approach apparently developed quite early and it was used in all public statements about the hunt. However, he felt there were problems inherent in the increased attention being paid to humane killing of seals in Canada and that the protest could be continued in spite of these efforts:

> I believe the humane societies were wrong when they went down this path of skinning alive because it almost developed then into a situation of 'If you stop skinning them alive, the hunt is OK and should go on. If you cannot stop skinning them alive the hunt should be stopped.' That really is not why people are against the seal hunt, in my opinion, Sir. They are against the seal hunt because it is brutal.... I have to be honest, and I think that the absolute essence of the hunt itself is what people are against; not any specific cruelty.

The above statement, given in testimony, was contrasted for those attending the hearings with a transcript of part of a WNEW radio news programme which had aired just previously in New York City: "Well, on the Canadian seal hunt in the spring of the year, large icebreaking vessels make their way to the floating ice, disgorge hundreds of hunters who then move into the breeding grounds of the harp seals, with clubs, and beat at the baby seals...with the intention of killing them but unfortunately for the baby seals the hunters are not always successful...in effect he skins them alive."

This writer once heard an early '70s seal protester describe the hunt. It was on a radio late-night talk show which invited listeners to call in with their questions and comments. One old lady called, sobbing, to say that she would gladly send all the money she could afford if only the speaker would use it to take those babies to a safe place, away from the hunters. The host was obviously upset with this reaction and tried to make her understand that such a "rescue" of individuals was not what the guest had in mind. But this poor old person would have given anything to Davies or anyone else who promised to end the hunt. Cruelty sells.

In the hearings, however, Davies admitted that his opinion of what constitutes cruel behaviour may justifiably be questioned:

> I go on to say, of course, that cruelty is highly subjective. It is cruel to me, but I recognize that it may not be cruel to other people.

When asked about his statements that seals are commonly cut open and actually skinned alive, Davies indicated that he understood the reflexive action which most clubbed animals display and that he knew it did not indicate consciousness:

> You have already had testimony or reports that suggest there is a great deal of post-mortem reflex action. The seal in the film that is wriggling with its stomach cut open — I have no idea whether it was alive or dead. In fact I have seen many seals moving like that that in fact are dead — their skull is crushed.

However, when asked "Have you seen baby seals skinned alive?" Davies replied:

> I have seen a seal with the hunter in the process of skinning it that I believe was alive at the time.
> *[Questioner:] One seal?*
> [Davies:] One seal.
> *[Questioner:] Out of how many taken that year?*
> [Davies:] Maybe 80,000.

In fairness, Davies also stated that it was his belief, although not his personal experience or knowledge, that there was evidence that other seals had been "skinned alive" because a veterinarian whom he had taken to the ice, Dr. Elizabeth Simpson, had examined seal carcases and she "found that a large number of skulls were intact, and in her evidence said that she believed these animals were alive when the hunter started to skin them."

Davies was asked why he had started a worldwide campaign to end the hunt on the basis of having seen one seal which he "thought" was alive at the time when its pelt was being removed. This would seem to be a fair question in light of his other statement about the post-mortem reflex movement.

Of course, it must be pointed out that while a crushed skull is an ideal way to prove that the seal was braindead before being skinned, the animal is also completely bled out first in order to insure a high pelt quality. Since exanguination takes fifteen seconds, it is easily conceivable that a seal could be completely stunned without sustaining a fractured skull and be bled while still unconscious. The skull might not be thoroughly crushed, although in these young animals (average ten days old), such damage is easily achieved. Even as early as 1965 stunning and bleeding out was mandatory, and had been routinely done by all sealers anyway for practical reasons.

Davies has always made a point of taking the uninformed out to the ice for the show: people always react violently to the reflexes and the gushing blood of dead seals. His credibility has been built upon that old adage, "Seeing is believing." All have reported on what they saw and believed.

When considering all the statements which Davies made during his testimony, or which were quoted from other sources during the hearings, it becomes obvious that he knows public sentiment and has used those elements which would be most effective.

During the hearings, Davies also stated that he knew harp seals were not endangered or in a depleted state. Charges that the seals may become extinct because of the hunt have been made by other organizations, but Davies' IFAW has concentrated on cruelty as a selling point. Perhaps he considered the extinction threat one which would easily be countered by the Canadian government, or he may have felt it would mean less to his donors than cruelty to individual animals.

The IFAW sends six newsletters to its membership each year. These routinely describe seal slaughter, and pictures of seal pups writhing in blood have been a mainstay of the campaign for more funds with which to stop the hunt. A favourite attention-getter has been inclusion of a "black death" folder of photographs enclosed with an appeal for funds and letter-writing to officials. As a marketing device, this is sensational. It cautions the reader not to open it unless s/he is prepared for the worst.

Two versions have been sent out since 1982. The earlier one has a message on the front which invites one to open it only after having read the enclosed separate message. The one sent out in 1983 has identical photographs and design inside and out, but a different message: now the word is that Canadian fishermen are the ones doing the deed. This is meant to support a proposed boycott of Canadian fish products.

One photograph shows two pups which have been clubbed: one is still undergoing a reflexive action; the other one has been clubbed and bled out, yet the eyes are still open. Therefore, it appears that the animal on the right is still alive and conscious.

The same shock is achieved with another photograph. This pup has been either shot from underneath (blood is gushing *up*) or impaled with a hakapik, which is not in either case a typical, practical or legal method of killing whitecoats. The great amount of blood near this animal is an indication that it is dead. The eyes are still open and anyone who does not know about sealing could assume that the seal was still conscious and was suffering.

A third photograph also tends to convey the impression that the pup shown is alive and suffering, having been clubbed and bled, yet *still* moving. This scene takes advantage of the average person's lack of knowledge of the reflex action, the "chicken with its head cut off" movement. Such movement is not an indication that the animal felt anything while it was thus moving. Such side-to-side swimming motion is, in fact, a usual sign of braindeath in seals.

The last photograph is most distressing. A whitecoat pup has been clubbed and then bled out. An adult harp (which could be of either sex) has come up to this body, which may or may not be that of her(?) pup. A reasonable explanation of this scene is that the adult has been at that point approached by a human, raises its head in the typical alarm position of the harp seal, and snarls. Within an instant, it will turn and lunge off to the exit hole in the ice.

However, to the person who does not recognize this posture, or know this behaviour, it would certainly appear that this adult is a female, keening her grief over her slain offspring. Male harps as well as females may on occasion come up onto the ice to rest after pups have been slaughtered. Adults have been seen to sniffle at pup carcases, although it has not been confirmed (as Davies has stated) that dams ever "try to nurse the shattered remains."

It should be pointed out that pups which have been clubbed and then bled are immediately skinned, thus removing any identifying scent to which a mother seal might be attracted. They are not left on the ice to cool down. It is to the workers' advantage to remove the pelt immediately before it can "ice burn" from the heat retained in the body.

The photograph of the adult and pup was taken on a bright sunny day. An experienced sealer would not have left this pup and gone greedily on to others; he would have pelted it immediately, since the bright sun would have been a compelling reason to quickly remove the pelt.

Since Davies has often used photographs of bloody seals in his fund-raising and "program services" literature, he must feel the tactic is a strong selling point for credibility and sympathy. Although it has been muttered that at least some of Davies' photographs and movies appear to have been staged, this is not only difficult to prove, but probably unnecessary in any analysis of reasons for his credibility.

Although some of the black death photos appear to have been taken of actual hunt scenes, and others look as though they may have been "arranged," the effect is the same. A humane person is inevitably terribly upset by the sight of any of them.

This person would be equally distraught at watching the seal hunt in person, or by a visit to a slaughter house. Hogs, cattle, sheep and poultry also die bloody, wide-eyed deaths and exhibit some extreme reflexes. No sane person would enjoy seeing them. Thus, although in the case of the black death folder three out of four scenes are extremely upsetting, and two of them (the one of a pup with a blood-spouting hole in its head and the one with the "mother" and pup) could have been staged, the total effect is one of shock and belief in the accompanying message: "Please help IFAW stop these atrocities."

Such scenes have literally been worth millions of dollars in donations to this organization. The money has been poured into an international flood of advertising against the hunt, which has asked for more funding to stop these atrocities and to end this cruel bloodbath forever. This tactic has worked well since the IFAW was established in 1969 as a fund-raising protest organization.

It has been speculated that the International Fund for Animal Welfare may have been formed due to internal problems Davies was having with others in the New Brunswick humane society over his depiction of the seal hunt. During the Fisheries and Forestry Committee hearings in 1969, it was disclosed that the managing director of the Saint John, New Brunswick SPCA had publically disputed Davies' claim that seals are commonly skinned alive. He offered $1000 to anyone who could prove it is possible to skin a live seal. At the time of the hearing, no one had taken him up on his offer.

There must have been some very heated discussions in that province of Canada over this whole affair of Davies' campaign against local fishermen who become sealers each spring. For a humane society official to make such an offer is unthinkable; suppose someone might try to commit such an act! But no one did. Sealers know better, and who else would try?

The rhetoric has continued to flow and has to be repeated here for the emphasis it deserves as an example of a successful psychological pitch (February 1982):

> IFAW wants to end the Canadian seal massacre entirely — no half measures — no compromises. Just an end to the slaughter. Here is how we plan to do it: IFAW will confront the killers on the ice floes this winter to keep the facts of this massacre before a world audience.
>
> IFAW and I must help these suffering creatures, but we're totally dependent on you and other good people like you. Thankfully, like most of us, you believe in saving animals — a cause that is noble and good.
>
> So please, help me to keep on helping all of these animals while at the same time continuing IFAW's determined efforts to save the seals — animals tortured beyond belief and needing so badly our pity and action. Every spring, more of them die horribly. So the sooner I hear from you, the sooner we can help the helpless victims of springs yet to come. I beg you to open your heart now.
>
> Yours sincerely,
> Brian D. Davies
> Executive Director

This short but emotional appeal is only a small excerpt from that particular newsletter from Davies to his members. It goes on to detail the opening of offices in Canada and Belgium where lobbying would be carried out. Davies has been accused of many things, but never of thinking small.

The seal sanctuary and tourist attraction in the Gulf of St. Lawrence never materialized. The hunt continued to take place there despite Davies' pitch for replacing it with sight-seeing tours. Legislators apparently did not believe that local people would benefit particularly from it, and had no faith that they would care to give up hunting seals, since they saw no reason why they should stop.

When one considers the bleak solitude of the Magdalen Islands, which are wind-swept dunes and rocks, the prospect of luring thousands of people there for any period of time longer than an hour and a half, feeding and bedding them down, and handling their waste disposal, is rather preposterous. It would disrupt an entire community's way of life which is based on fishing and sealing and not necessarily improve it, either materially or spiritually.

Once he decided that he could not talk the Canadian government into accepting the sanctuary idea, Davies began to make public claims that he would end all sealing in the Gulf, on the Front, and in native communities. Despite his former statements that specific cruel acts were not so important to the public as is the general idea of baby seal massacre, his messages continued to stress "horrible torture," "skinning alive," and the long, lonely hours that grieving mother seals spend weeping after they have found their pups reduced to shattered carcasses.

Apparently, specificity doesn't hurt fund raising, or he would perhaps have stopped. This has obviously been a formula for making money which is recycled into more ads which bring in more money, ad infinitum.

By the 1980s it came down to a simple, single theme: the seal hunt is morally wrong from any animal lover's perspective, and for this reason must be stopped. Claims of cruelty continued to be used in support of the stop-the-hunt-at-all-costs demand and became justification for the proposed 1983-84 boycott of Canadian fish products.

Canadian fishermen are the ones who have traditionally been seal killers. They feel strongly that the "seal fishery," as it is called in Newfoundland, is as much a traditional and natural part of the annual cycle as is the lobster fishery, the cod fishery, the herring, caplin, turbot, mackerel, and so forth.

Fishermen across Canada have made public statements to the effect that they support the east coast fishermen in urging the government to continue support of the hunt and not give in to the pressure of protest. This support has been bolstered with figures pertaining to the years of seal management which have reportedly resulted in herd increases and in humane regulations of the whitecoat hunt itself. Fishermen all realize that the year has resource gaps, when there is no income to be had. The seals fill one such gap for the east coast fishermen. They form a vital link in their economy.

The fact that Davies himself has recognized this at least since the late '60s does not seem to matter any more. By 1982 he had accomplished the ruining of the European sealskin market through his multimillion dollar campaign against the hunt in Europe. The Common Market Parliament, bowing to the pressure of several million names on a petition, ended up recommending a ban on whitecoat and blueback (hooded seal) pup pelts in their countries.

The IFAW campaign against the seal hunt now had to demand an all-out ban if it was going to keep its credibility. For years ads had been stating that, with enough public donations, the cruel hunt could be entirely ended forever. Enough bait had been cut, it was time to fish in earnest.

Davies had discovered that the hunt would not be stopped by attempts to influence or shame the consumer of fur products into not buying them, thus influencing natural marketing. Consumers who wanted furs, sealskin or not, bought them. Another problem with that approach was the fact that sealskins were very often rendered unidentifiable.

Although protesters all claim that the pelts are used for luxury furs and for trim, a whitecoat pelt, dyed brown or black to further hide its origin, is not expensive. The fur is longer than mink, for instance, and is not as durable through long wear as are many other furs. Many of these pelts are turned to leather and used in shoes, vests, pocketbooks, and slippers. It would be impossible to influence the average consumer against buying a leather item which does not resemble a seal in any way, which could just as well be lambsuede, or deersuede, or kid goatskin, all of which have a social tradition of acceptability.

Perhaps this is one reason why Davies sought a legislative bypass of consumer wishes in order to accomplish an end to the seal hunt. He lobbied extensively in Europe, especially in France and West Germany where the leather industry is very strong and where furs have long been made into finished products. The German magazine *Der Spiegel*, analogous in function to the United States *Time* magazine, carried an article in April 1983 on Die Robben-Schlacht (The Seal Slaughter). West Germans had been treated to such an intense campaign that their politicians leapt into it in order to gain public favour and to enhance their images as seal saviors. Politics was to become an important lever in the seal wars.

In France, Brigitte Bardot had been carrying on an emotional campaign with Franz Weber who had once offered to set Newfoundland up with a fake fur factory if sealers would give up killing seals and stuff toy animals instead. Although Weber's plan had been rejected soundly by Newfoundlanders, he had continued in the animal rights movement with his Switzerland-based United Animal Nations. His activities in France and the rest of western Europe had, in general, enhanced Davies' anti-seal hunt campaign.

Der Spiegel pictured Bardot and Weber at a news conference, sitting with stuffed seals before their microphones. The public was enthralled by this beautiful woman claiming to have cuddled a live seal out on the ice and, since that time, devoting herself to saving them all from the cruel men who hunt them. Millions felt empathy with her recollections of this emotional experience.

According to a recorded Newfoundland radio broadcast, Ms Bardot had missed going to the ice with Davies on the day they were to have met at St. Anthony Newfoundland. Her plane was late and he had been having significant stress with local crowd trouble. He finally took off without her and she had to hitch a ride out to the seal ice with a Greenpeace group. Once there, she reportedly had difficulty due to a very cold lack of bathroom facilities and left soon after. Allegedly, she was photographed at some other time, hugging a stuffed seal.

Thanks to Bardot, Weber, Greenpeace and Brian Davies, the campaign against the seal hunt was a very familiar topic for Europeans. By 1983 European fur dealers had decided to shy away from offering sealskin in any form. A few terrorists had also convinced fur shop owners not to carry any sealskin by breaking windows and putting glue in door locks. The result was Davies' long-planned goal: all forms of sealskin were either off-limits from a legislative standpoint or were socially and financially ostracised.

Sealskin stocks in Europe were either disguised, put into storage, or shipped out to the Orient where they might be used as leather in anything from sneakers to jackets and pocketbooks. No new sealskin could be imported into Europe now, and the effect was a nearly total cessation of the commercial trade in any seals.

One might ask why the campaign of the International Fund For Animal Welfare did not then come to a close. The whitecoat hunt, which Davies had fought for years, was effectively over. The market for older "hair" sealskin was ruined everywhere in response to European speculators' fear of being stuck with a product which was legally imported but socially shunned.

A cynic might reply that if the IFAW had admitted to the public the actual scope of its victory for seals, then the money might cease to flow since the job had obviously been done. This, then, would be a "reason" why Davies continued to advertise that seals were still being brutally killed and that the older ones must be saved also.

In fact, the *New York Times*, Sunday, March 20, 1983 ran an ad which told of a partial victory for seals, thanks to Davies' successful European campaign. In March of 1983, however, the mass slaughter of older seals which he claimed was about to begin, actually amounted to only 56,000. It included "tanners" (very late whitecoats), beaters, and older harps. While 56,000 seals may qualify as a mass slaughter of animals in the minds of some people, it certainly doesn't compare to the normal harvest of over 156,000 animals taken from the Front and the Gulf the year before the European pelt ban.

A reasonable explanation for Davies' continuation of the protest effort, and his 1983-84 attempt to blackmail Canada's fishing industry, has to take into account not only the continued financial welfare of the IFAW, but also his apparent obsession with actually ending the hunt. His international reputation depended on this: he had been stating for years that it was his goal, and to back off just when a "partial victory" had been won would have been out of character.

There is no reason to believe that Brian Davies could have been satisfied with an end to whitecoat slaughter only. He had been proclaiming his love of all seals for too long. No one who has followed closely his twenty-year history of involvement could really believe that he was motivated only by a desire to keep the millions of dollars rolling into IFAW bank accounts, even though that wealth was certainly a sign of his effect on the people who believed in him.

Not a model of honest reporting, Davies' lurid, maudlin newsletters about seal death and suffering appear to have been coldly calculated to wrench funds from anyone who can read and interpret a photograph and who has a heart for animals. Yet, if that is what it takes to accomplish his goal of an end to the hunt, then perhaps he justifies the "schmaltz" as a necessary means to this noble end. Certainly, the posture of an all-out life effort to save the seals will have enhanced his self-image, and support of the effort has probably done the same for his supporters.

There is in this case a similarity with certain religious leaders who have achieved tremendous support through both their charisma and acceptance of their doctrines. After that achievement, one does not simply step down when goals of belief have been demonstrated; one continues to reiterate those goals until they are all finally realized. In the case of the hunting of western Atlantic seals, it is not likely that a complete cessation shall ever be accomplished, for various reasons.

Perhaps this is why Brian Davies and other protest groups and their leaders are continuing public campaigns and fund-raising drives to further seal hunt protest in a time when sealing has just about come to an end. Surely the money has to continue to be donated in order for each group to continue. And just as surely, there shall be other deserving wild creatures which should be saved in the future: whales, dolphins, sea turtles, kangaroos, Philippine dogs and South Korean cats are all in danger of cruel treatment by man. If the public is willing to continue to "fight" this with its donations, then why should any humane or environmental organization leave the arena? It is not likely that any shall, for there is much more money to be gained in the business of humane and environmental protest. As a lawyer for a now-defunct sealing company recently said, protest is a growth industry.

Signs of progress, such as the greatly decreased take of seals in 1983 and 1984, have been triumphantly announced by the IFAW, Greenpeace, Paul Watson, and anyone else who wants the public to believe that its support has led to a decreased kill. However, lack of a market was not the only reason for a small catch.

The weather and ice conditions in 1983 and 1984 were absolutely terrible for both the Gulf and the Front. When sealing season arrived, the ice stretched for two hundred miles offshore from eastern Newfoundland, and harbours were blocked tight. When inshore ice finally broke up, about the time that beaters would have been normally ready for taking, winds came up from the west and blew shore ice out against the arctic pack which held the seals. The jam was so serious that even large ships could not break through and longliners found it nearly impossible to sight any beaters.

Young seals did not travel across or under the great expanse of jammed ice from shore to reach its western edge in any great numbers. Longliners which, in a good year, might come home with as many as one thousand animals, were lucky to find three hundred. Meat prices rose, as demand far exceeded the supply, but pelt prices were cut in half since the buyer (Carino Company at Dildo, Newfoundland) had stated that it would take only 60,000 pelts, and those only at half price in case they could not be sold.

As a result, fewer men risked going out in those two years. Fuel was expensive, shells were expensive, men out sealing had to be well supplied with food for weeks on end, and the chance for financial gain was small. Some hardy souls decided to try their luck at lower cost by going out to the ice in their outboards for a day at a time. Each "speedboat" would roar across twenty miles of open heaving seas to the ice edge, hunt for a day, and then try to return. Each boat rode dangerously low with the few seals piled in the middle. The return trip through heavy waves and increasing evening winds was extremely hazardous.

At St. Anthony, on the northern-most tip of Newfoundland, a number of small boats became jammed up in heavy ice and had to wait for a government ice-breaker to get in to free them. The winds had moved the ice and blocked their escape. Although all were saved, it was a trying experience. Some, however, went out again afterwards. Newfoundlanders are persistent as a rule; they hate to admit that they can't do something.

Even if the weather and ice conditions had been ideal, however, fewer fishermen would have been able to depend on sealing in '83 and '84 to tide them over to another ice-free season because of the limited number of pelts which Carino had said it would take from them and because of the drastic reduction in price per pelt.

Of course, Newfoundland and Quebec citizens were not the only ones affected by the loss of a ready market for sealskin. Whole communities of native people, both in Canada and in Greenland, who had always traded seal pelts for staple foods, guns, ammunition, clothing, medicines, paraffin to heat their homes, flour, chocolate, and gasoline for their outboards and snowmobiles, were suddenly deprived of this means of supporting themselves. They could no longer use sealskin for currency or as a trade item. It didn't matter that the older seals which they hunted were not covered by the official ban in Europe, the unofficial reluctance of far away buyers to accept any seal products precluded continuation of even this small part of the market.

Brian Davies had won, even by 1983, in the sense that his advertisements had generated an official pelt import ban which, in ripple effect, was extended to all products which have the label "seal" attached to them.

In the modern Arctic, the end result of Brian Davies' humane new world has been hunger, cold homes, no new clothing, less gasoline for essential travel, and less ammunition in a time when the old ways, before guns, have been largely forgotten. Now that sealskin no longer works to bring in outside goods and food, the only alternative is welfare subsidy for many people who had formerly been entirely independent. The subsidy falls far short of the value which seal pelts used to bring.

Throughout coastal eastern Canada and Greenland, everything has been changed. A very real human misery and deprivation is the result of the many millions of dollars donated by good Americans, British, French, Germans, Swiss, and other Europeans to save harp seals.

Still, Brian Davies continued his plea: "But the task of saving the seals is only half done. Please open your heart today. Join a winning team to save the older seals and stop the hunt forever." Under a picture of a large harp, the message reads: "Please help it's VITAL. I want to save the seals from any more torture. Please accept the donation indicated below to suport your anti-seal-hunt campaign. To IFAW. Here is my contribution in the amount of $...." (*New York Times*, Sunday, March 20, 1983).

Perhaps Davies would never tell them that they had already done a good job. Surely he would never admit to his public that IFAW action had had a severe economic effect on anyone. The IFAW position had always been that the hunting of seals was "damned unnecessary," and that the seal hunt would readily be given up if another source of income was available. "It's only a matter of dollars and cents," had been Davies' reply to the economic issue. His public posture was that the Canadian government should outlaw the hunt and compensate sealers for the lack of this resource opportunity.

This would perhaps be in the form of extended UIC (Unemployment Insurance Compensation), which many fishermen accept in the winter anyway. The plan is identical to that proposed by Greenpeace and is rejected in principle by sealers everywhere. They know that such payments would not substitute adequately because, in dollars and cents, UIC is less than half the potential to be made from sealing in a good year.

It was obvious that the government still supported sealing. In 1983 the price for pelts had been cut in half at the Carino plant and there was a call from sealers for price supports to make up for this loss. After long debate and much newspaper attention, the sealers were promised price support on the pelts they had sold. Each man would receive a cheque to compensate for eighty percent of his loss due to the lower price. The total plan amounted to about one million dollars paid in price support, although it came very late after the '83 season.

Further government aid to sealing was seen in continued support of the Canadian Sealers' Association; this group's expenses and staff were federally funded, as were their feasibility studies into teaching local people how to process and work with their own seal pelts. Crafts projects were thus "encouraged" and it was reported that some efforts were being made to find foreign markets in the Far East for sealskin products. A shoe factory began to make sealskin boots, and vests were also produced. It became evident that a rather surprising Canadian market for sealskin might easily be developed.

Brian Davies must have been aware of all of this, and probably realized that although these people had not had a tradition of using the products made from their seals for many years, another form of social change was taking place. Newfoundlanders, in furious response to years of being labeled brutes for killing seals, had decided to wear the buggers themselves if no one else would.

The number of seal products which were made and sold did not represent very many animals by anyone's standards. However, those few products were the start of a new hope for Newfoundlanders to repair their damaged economy and regain their sense of pride and accomplishment. By the November '83 Marine Show in St. John's, Newfoundland the public mood was right. Four thousand dollars worth of sealskin was sold to the public in the form of boots and vests. A small beginning had been recognized, and by February 1984 hopes for federal grants for craftspeople were being cautiously expressed.

Local men and women in small fishing villages all up and down the east coast were thinking about learning to sew up practical things, such as boots and mittens, coats and vests. They knew that it was wrong to give up on sealing just because outsiders thought they should. And they had confidence that a cross-Canada market for their products was possible.

At a time when it seemed that everyone had hit bottom due to the IFAW campaigns, and that the hunt was finished forever, the old resilience which had always kept them going, mending the nets and returning to the sea, surfaced once again. It was perhaps no coincidence that Davies decided to keep on advertising to save older seals. Perhaps news of this resurgence of interest in seal products in Newfoundland was a good reason to continue the campaign. Seals were still being killed for their skins, and though the European market could never be equalled, there was still economic incentive to continue the hunt.

Analysis of tactics used by the IFAW, however, reveals that another anti-hunt scheme had already been conceived and set in motion before this time. It was formally announced in July of 1982 that a campaign to boycott Canadian fish would be instituted in Europe and the United States. It was claimed that this boycott would exert severe economic pressure on the Canadian fishing industry and that this pressure would be ruthlessly applied until the government officially banned the hunting of seals.

The European pelt ban had not been one hundred percent effective in stopping the hunt, although from a species-impact perspective, if it continued for any length of time it certainly could have long-term effects in altering the life conditions for seals, whales, sea birds, fish and shrimp in the western Atlantic. However, a species impact was apparently not what Davies had in mind. His focus was a demonstrable end to the killing of *any* number of individual animals which were "clubbed-drowned-shot...for money."

Although his victory in Europe had set a historic precedent which could have lasting effect, the ideal goal had obviously not yet been realized. The fish boycott scheme, which called for an *official* and permanent end to the seal hunt, was now advertised as the only practical means to this end. It would be the ultimate demonstration of the power of the International Fund For Animal Welfare, and a tribute to the leadership of its founder.

At this point the protest movement had accomplished a number of goals. Those organizations which had had no effect at all on the European pelt ban nevertheless gained credibility through it as part of a successful movement to curtail the taking of a form of wildlife. The repeated annual publicity generated by the IFAW, Greenpeace and, occasionally, other groups, on the ice and in urban demonstrations on the steps of capitol buildings, had all impressed the public with their ideals and their power to attract media attention. Individual supporters felt increased loyalty and faith in their own animal rights organizations and renewed their annual memberships regularly.

The seal campaign success in Europe was a lesson in tactics which would never be forgotten. It had been won with millions of donated dollars poured into newspaper and television advertising across the continent. These ads generated petitions bearing several million names, adults and children alike, calling for the ban. Representatives to the Common Market Parliament were afraid to ignore this noise from the constituency. Political power was being exerted by those who vote and candidates knew they had to demonstrate compliance with the electorate's wishes.

At the time, the seal issue created a needed diversion from the pressing economic problems felt in each nation. It served as an emotional outlet and as a symbol of control in a time when financial hardship on individuals and communities alike tended to convey a general and increasing feeling of helplessness.

It made people feel good to contribute or sign the petition. In this way they could make affordable positive statements about a problem which had been clearly outlined for them and for which there was promised a single solution; the ban would end the seal hunt, and the innocent babies would be saved through small sacrifices by many good people.

Although Canada tried desperately to exert influence at the diplomatic level, and although ministers of state across Europe quietly shook their heads and agreed that scientific assessments of the seal issue were correct, they had to agree that they also were moot. They knew that the public would never be dissuaded from the "Save The Seals" perspective. The $300,000 which Canada poured into counter advertising was too little, too late, and could never hope to overcome the previous years of increasingly distressing IFAW propaganda.

At the time of the vote by the European Common Market Parliament the actual count was resoundingly in favour of a ban on harp and hooded seal pup pelts. However, the figures should be examined closely in order to fairly assess the effect which the IFAW had produced.

There were at the time 410 members of parliament. On the day of the vote, it was noted that twenty-eight percent were absent. Less than forty percent of the parliament voted to ban; 126 representatives were present on the day of the vote and abstained; 10 members voted *not* to ban. The final tally is as follows:

- 114 — absent
- 126 — present, abstained or did not vote
- 10 — present, voted *NOT* to ban pup pelts
- 250 — (or 61 percent) did not vote to ban for one reason or another
- 160 — (or less than 40 percent) voted to ban the pelts

Although Davies claimed to have had tears in his eyes when the votes were counted, claiming a victory of one hundred and sixty to ten, one should not forget the above. It perhaps reflects representative dissatisfaction with what they perceived as unfair pressure brought to bear upon them.

The Council of Ministers called for a voluntary two-year ban on the importation of whitecoat or blueback pup pelts into all member nations. The ban was to begin in the spring of 1983 and the matter might be brought up for official discussion again in October of 1985. There was thus a two-year hiatus of heavy commercial exploitation of seals, at the same time as a period of heavy ice each spring contributed to a decreased take.

Some whitecoats were still taken by those who cared to

demonstrate defiance of protest. The meat was consumed as usual and the pelts were stored. In 1983, although it had been announced that no whitecoats would be accepted at the hide plant in Dildo, Newfoundland, it was reported that some five thousand pelts officially classified by the Department of Fisheries as "whitecoats" were indeed purchased. They were late tanners, not yet into the raggedy jacket or beater stage of pelage. The fact became a small media victory for Greenpeace and IFAW, and one more public justification for pushing the now-infamous fish boycott idea.

In retrospect, the disclosure that some "whitecoat babies" were still being killed by the cruel fishermen, even after the European ban on their hides, was perhaps the only thing which the media could use as "news." Since ten-day-old white pups were no longer being clubbed en masse and Greenpeace and IFAW were no longer able to stage large on-ice demonstrations there was very little to attract journalistic interest. Greenpeace USA spokesman Peter Dyskstra jubilantly gave the news of continued whitecoat slaughter on a New York City radio talk show in the spring of 1984 and used it as an example of Canadian "double talk" about the realities of a reduced hunt.

The only other newsworthy events which were carried by the media were occasional arrests of Greenpeace members flying too low over the ice during the hunt and the fact that a group of Magdalen Islanders, fed up with the IFAW, turned over an organization helicopter when it was in for refueling. It sustained some significant structural damage (which was allegedly exaggerated for the press) and was subsequently repaired and reported to have been flown across Canada in a fundraising demonstration which some sources said netted more than the damage had cost. But this was not particularly newsworthy, either.

By the 1984 sealing season harp seals had become a dead issue except in local areas of eastern Canada where a little residual protest action still hung on. Only short term, low-level media interest was generated in a much reduced hunt, mostly with rifles, for older, less vulnerable, less attractive animals.

The demise of the large commercial hunt was obviously recognized early on by the IFAW as a problem which would cause faltering media coverage. Without significant press disclosure of the plight of baby seals (due to a reduced hunt), it would be next to impossible to convince the international public to boycott Canadian fish. Since the hunt would not be covered voluntarily by the press as news, the boycott might have to be promoted through paid ads unless the media could be convinced that the boycott itself was newsworthy.

Not only those who make decisions about such things, but the housewife, ultimate consumer of fish products, would have to be convinced that the boycott was important and necessary in a continuing humane effort. According to Steve Best, Canadian coordinator for the IFAW, the link in the mind of the housewife between baby seals being killed and any fish products in the market would have to be "dog simple" or it would be ineffective (personal telephone conversation with author, October 1982).

The effort to promote the boycott had begun early in 1982 with advance notice of the plan being given to the Canadian Department of Fisheries and to major British and American fish importers who used Canadian products. The public would hear about the scheme after the initial stages had been set with those in power.

The Seal Wars were on again. The fish boycott idea was finally brought to the English public in November 1983 with the announcement of a press conference at which plans to blackmail major importers of Canadian fish would be disclosed. All representatives of major newspapers and television stations were invited for drinks, lunch and new photographs of seals actually being drowned in nets. The announcement included the news that the latter would be of special interest to women. It did not say why.

According to reports, however, the news conference was a flop. Major media representatives did not attend, atlhough plans for the fish boycott were announced there anyway. Perhaps this media snub was a signal that they would not be so easily used this time. The long-awaited and much-feared boycott of Canadian fish had started in England, but the drum roll was much subdued.

The original fish boycott plan had not been meant for the eyes of the public, but was intended to be a scare tactic of significant proportion. The "European Economic Community, Canadian Fish Products Boycott," a paper "prepared by Stephen Best, Co-ordinator, The International Fund For Animal Welfare (Canada)," was sent to major buyers in order that they fully understand what was coming and so that they would apply pressure to the Canadian government prior to the spring of l983. The pressure to be applied was insistence by them that the seal hunt be officially abolished.

Although Canadian Department of Fisheries and Oceans personnel were very upset by the potential outlined in this paper, the hunt went on as scheduled, albeit much reduced due to the European pelt ban. There had not been any social or political precedent for calling it off and the boycott scheme was regarded with both fear and confusion.

Of course, no one knew just how much damage it could do to Canadian fishing interests, yet it was felt that if the government gave in to this sort of blackmail on the seal hunt, it would be even more vulnerable the next time a special interest group decided to apply economic blackmail to an industry.

Internal turmoil, however, was spectacular. The respective bureaucracies of the Department of Fisheries and Oceans and of the Department of External Affairs were hotly engaged in debate over how to proceed in handling the inquiries of those who had received the boycott paper. The question was one of posture: would Canada stand firm on her management plan or cave in to the IFAW in the face of enormous loss of income for her already troubled fishing industry?

The advance paper stated that the boycott would be launched "if there was a 1983 seal hunt." Actions to be taken were outlined as follows:

1. FULL PAGE ADS in all major newspapers explaining the boycott and identifying, where possible, specific brands and products using Canadian fish (a sample of the ads are included with this report).
2. ORGANIZED PICKETING of large retail outlets selling Canadian fish products.
3. A MASS DIRECT MAIL CAMPAIGN of postcards (one side printed with the same design as the ads), pre-printed with the mailing addresses of major Canadian fish exporters, to be mailed by consumers and containing their signed pledge NOT to buy Canadian fish products.
4. A PUBLIC RELATIONS CAMPAIGN prepared to respond to and capitalize on the inevitable interest this type of action draws from the various news media.
5. CELEBRITY ENDORSEMENTS of the boycott.
6. IN-STORE INFORMATION TEAMS identifying Canadian fish products for consumers.
7. Direct contact with British and European fish importers to explain the scope of the proposed boycott.

   An initial sum of 1 million pounds sterling is available to implement the boycott. It is expected, also, that the program will be financially self supporting with funds raised from the response coupons included with the 'print' ad campaign.

The paper outlined the expected effects of the boycott and the future of it as an economic force in the United Kingdom, Europe and the United States. It was claimed that hundreds of millions of dollars would be irretrievably lost by Canadian fish exporters. Business would never recover since, in the interim, other fishing interests in Iceland, Norway and Japan, the United States and the Soviet Union would probably have won suppliers' contracts and Canada would be permanently out in the cold.

The hype inherent in the scheme was further elaborated with claims that if the seal hunt continued, other organizations would probably also "introduce their own Canadian fish products boycotts," because such would be attractive fund raisers for each organization and the idea was a good way to attract and hold membership.

As if the above wasn't enough, Steve Best also included the news that the boycott would "be linked to public health concerns" in order to

> ...negate, for example, the proposed one million pounds sterling campaign by B.D.H. & Partners of Manchester to re-establish the canned salmon market in the U.K., hurt by the botulism scare in February of this year.
>
> Furthermore, public health concerns will also focus on the sometimes questionable quality of frozen fish from the east coast of Canada. Buyers are currently purchasing samples for analysis.
>
> There is no doubt that this tactic will spread to the U.S.A., where it could do considerable harm to Canadian fish sales, particularly to fast food outlets, where quality of product is an essential marketing tool.

As if all of the above was not implied threat enough, Best added that dock workers in the United Kingdom were ready to refuse to handle any seal or fish products from Canada because of the influence of Terry Duffy, a union leader who had "an outstanding track record as an animal rights activist within the trade union movement." The dockers in question were said to number "a staggering 12 million." Duffy reportedly was organizing to "extend the UK 'blacking' of Canadian seal and fish products to other EEC ports...it is difficult to judge how effective this action will be at the dock level although it was effective on whale products."

Best's paper claimed that such widespread action by the dock workers would encourage "extensive media coverage...reinforce the public's awareness of the proposed boycott of fish and the public health concerns about botulism and canned salmon." Canned salmon is a favourite British food item, and British Columbia has traditionally been a major supplier.

Unfortunately for the IFAW boycott plan, this latter threat to ruin the canned salmon part of the Canadian fishing industry was effectively foiled by the environmental effects of "El Nino," a unique natural ocean-warming effect which had been occurring off the Pacific coast of North America. El Nino had abruptly diminished the take of nearly all salmon by Canadian and American fishermen. The supply became so limited that hardly any new salmon was available in the stores in 1983 and 1984, and British fish lovers demonstrated that they would buy it from anyone. By late 1983 it was also evident that no dockworkers' action had had any real effect on the import of Canadian fish into the British Isles. That particular threat had not materialized. Maybe dock workers missed their salmon also.

On the face of it, a widespread boycott of Canadian fish would have been a logical, perhaps final step in an anti-hunting protest against the taking of harp and hooded seals by Canadian fishermen. The logic is simple: all fishermen across Canada had made public statements of support to the effect that they, through their unions or native organizations, would continue to urge the government not to abolish the seal hunt which was so desperately needed by their colleagues on the east coast. For this reason the IFAW boycott was to include all Canadian fish products, to hurt all producers so that they would cease to urge such government support of the hunt.

The Canadian fishing industry is gigantic. Sales to Great Britain amount to 115 million dollars a year, but Great Britain receives only about ten percent of Canada's fish; eighty percent is exported to the United States. If the boycott was indeed successful for several to many years, the orders cancelled could amount to a multibillion dollar loss. If this even began to happen, it is certainly conceivable that the Canadian government would be so pressured by the unions, primary producers and exporters that it would have no choice; the seal hunt would have to be sacrificed to save the fishing industry.

As Steve Best had pointed out, the seal fishery brings in only a tiny fraction of the gross national product realized by fishing. There is no question but that real economic pressure would result in capitulation.

The Canadian government did not officially abolish the '83 or '84 seal hunt. The fish boycott attempt in England proceeded according to schedule. In 1983 and 1984 a sample mailing from the IFAW reached 5.5 million homes. (The population of the United Kingdom is about fifty-six million.) Almost immediately Tesco Corporation, which was named as one company which imports Canadian fish, received some 10,000 protest cards from consumers who agreed with the stance of the IFAW. Other "target" companies also began receiving the pre-printed protest cards which had been mailed to the public by the IFAW.

Each card contained a picture of a man swinging a bat at a whitecoat seal pup: "Boycott Canadian Fish While Canadian Fishermen

Do This!" The message was "dog simple" and the media did pay it some attention with "man on the street" interviews and general news treatment. Tesco quickly announced that it would sell its inventory of Canadian fish and not order more while the seal hunt was still taking place. Initial Canadian reaction to this announcement was panic. It was feared that other major buyers would follow suit.

With such a promising start one would think that nothing else would be necessary for realization of the IFAW plan and a quick official end to an industry which meant so little in the overall economic perspective of coastal Canada. What else would it take to convince those handling Canadian fish that they didn't have a chance against such a well-organized scheme? Although it seems to be a case of real overkill, further threats which were not published in the original paper were delivered: some major grocery chain executives began to receive postcards regarding the fish boycott at their homes. These were not, however, IFAW cards. Each had a scene of street violence on one side and "Greetings In Revolution from the Irish Republican Army" printed on the other, along with a written message to the effect that Canadian fish products should not be purchased while the seal hunt was still officially sanctioned. No British citizen could ignore the implications of this suggestion.

This bizarre tactic was immediately investigated by Scotland Yard so as to coordinate handwriting analysis and comparison with other such threats by the IRA into a meaningful intelligence report. IRA card recipients lodged their complaints with the Canadian High Commission in London.

Why on earth, one might ask, should the Irish Republican Army care about an ostensibly humane seal-saving campaign? Certainly its record of humane concern has not been obvious of late. The answer lies in the potentially destructive impact of the fish boycott on the Canadian fishing industry, on shipping, on diplomatic relationships, and on wholesale and retail business in both countries. A connection of surprisingly wider significance was accidentally discovered by electronic surveillance (telephone wire taps) in London. These taps, routine corporate espionage conducted by a large Canadian corporation (not a part of the fishing industry), disclosed the reason for IRA interest in aiding the fish boycott. The IRA was to be guided and funded by an agency of the Soviet Union in its own anti-seal hunt campaign.

This agency is dedicated to covert sabotage of normal business and governmental operation in the free world, and operates internationally. It funds such terrorist organizations as the Irish Republican Army and the Palestine Liberation Organization. It is not concerned with the means of achieving disruption, so long as disruption does occur. Terrorism has been a major tool in this process and has always been regarded as a nearly indomitable force, likely to strike at any time or place.

English executives who received hundreds of IRA postcards were terrorized by the implied threat. They did not dare to admit or discuss such a connection publically through fear of reprisal. Perhaps few of them had heard of the Soviet connection, but the IRA was violent death personified.

Given the seriousness of the matter, it is surprising that all major importers did not give in and announce that Canadian fish would no longer be included on their order sheets. But perhaps British stubbornness played a role in resistance. Only Tesco Corporation immediately announced that, due to having received some 10,000 IFAW cards from the public in one week, it would respect its customers' sympathies and reorder elsewhere. It is not known if Tesco decision-makers received IRA cards at their homes.

Other companies hedged and appeared to be looking to Canada for answers as to what to expect. A state of chaos had indeed been achieved for the time being, although no other firms announced that they would definitely not reorder Canadian.

As pointed out previously, the IFAW had sent its boycott plea to only 5.5 million British homes. While a deluge of 10,000 protest cards in the space of a week would be a shock to any corporation, the following facts have to be considered: the response *rate* to each company was actually very low, being somewhat less than .3 of one percent of the total homes contacted by the IFAW; further, there was the possibility that not every card would have represented a lost customer for that particular chain even though that customer did bother to send out the pre-printed card; finally, the British population of some fifty-six million people contains a high proportion of animal lovers who feel sympathy with the save-the-seals idea, but who do not care to give up their fish, a traditional food item.

There may be a basic similarity between the public response to the IFAW pelt ban campaign in Europe and the English response to its fish boycott. In each case an impressive number of people responded to the cruelty-focused advertisements. However, in each case the *percentage* of those who responded was very low considering the total population which had been contacted. The European ads were inserted in newspapers with an estimated readership of some forty-four million. The ads generated letters to elected representatives of the Common Market Parliament. These amounted to many thousands and the total names on petitions presented to the parliament numbered several millions.

In each case the public was visibly moved by the plea to save baby seals and in England there was a request to boycott products which were scarce and considered necessary and staple food items. Yet, when percentage of response is considered, the difference was not pronounced. It may be that animal lovers who will act on such a suggestion exist in fairly large numbers in all of western society, yet constitute a relatively small percentage of total population. If that significant segment of the total is reached, it will make the called-for response.

Total numbers reaching any one targeted corporation or member of a legislative body may be impressive, but may not constitute a fair sample of either potential customers or eligible voters. The result may be a highly-skewed illusion of reality and of the power behind any particular vested interest group.

This factor may have been known to the parties involved in the Soviet-IRA boycott connection, and could have been why the implied threat of terrorism was added to each corporation's collection of publically endorsed IFAW cards. It may have been seen as extra insurance that a major economic disruption would occur.

There are some who would like nothing better than to demonstrate a real connection between the animal rights movement and the USSR; the added bonus of a symbol of notorious terrorism such as the IRA, cooperating with seal saviours and the anti-fur movement generally, would be priceless. Unfortunately for those interests, however, there is no sound evidence that the IFAW had anything to do with the IRA in this or any other circumstance. The International Fund For Animal Welfare has long been dedicated to increasing its credibility with animal lovers through specific "humane" projects. Its activities have apparently brought in more money in donations than it has spent, and the organization has an affluent appearance. With this as a background, there is no reason to believe that it would want to have anything to do with international terrorism.

In England IFAW appeared to be doing very well with the fish boycott on its own. Animal lovers were responding in the expected percentages, at least initially. Large corporations were afraid of the influence that IFAW might have on a more general public. The press was finally beginning to take notice again and the hype about massive union involvement was generating widespread fear in Canada.

There is no reason to believe that the British arm of the IFAW would have wanted any involvement with the IRA. They certainly knew what disastrous results disclosure of such a connection would have in the press. Their public support would instantly disappear at the slightest suggestion of such a taint to the operation.

The only possible hint of a knowledgeable, purposeful connection between the IFAW fish boycott and an ultimate Soviet involvement is based on this writer's conversations with Steve Best. Best sent me a copy of the fish boycott paper in early October 1982 with a note asking me to call him with any comments or questions.

Topics discussed included his opinions on ways to effect specific social change. Best claimed that the animal rights movement was "parallelling exactly the writings of Karl Marx." This sounded like such malarky that I could not comment and felt great annoyance that he might think I could take him seriously. The topic was changed.

In retrospect it is apparent that the boycott plan contains the idea that the will of the people can overcome the greed and insensitivity of major capitalists who can be brought to their knees with economic pressure from the masses. If the boycott had worked, it would have been a great social experiment and a demonstration of the power of the animal rights movement which no politician could every forget. The power which the IFAW had demonstrated in Europe showed it to be an "extension of the people" organization. Ostensibly it owed nothing to any real power source, other than the public itself which it served and whose wishes it reflected. The public was obviously keeping it solvent.

Early in 1984 Stephen Best suddenly disappeared from the IFAW organization in Great Britain. One theory to explain this is that the "colonial upstart" had not been very popular with IFAW staff there due to his alleged abrasive and aggressive, overbearing attitude. Best had come in to "launch" his boycott plan but reportedly had been a source of concern and dissention since his arrival.

If British staff did discover or even suspect a connection between Best and any IRA involvement, it would have been natural to insist to Brian Davies that he be removed. This would help to defuse an investigation by Scotland Yard which would naturally follow as soon as the IRA cards were turned over to that agency for analysis.

Later in 1984 Best turned up again in Canada, serving the IFAW as a "consultant." Then, in March of '85, he participated in a panel on sealing at McGill University in Montreal, now as a representataive of the I Kare Wildlife Coalition.

Best's name is on the advance plan outline which had been sent to government and business, and therefore it can be assumed that he composed that draft. It seems unlikely that Davies would have conceived the plan and then let Best have the credit in order to separate himself from the unpleasant connotations of blackmail and economic ruin because Davies' newsletters and advertisements all endorsed the plan and encouraged the public to support it. If one assumed that the boycott was Best's brainstorm, it may be reasonable to suppose that he would not have discouraged any forces which might have contributed substantially to its success. But there is no evidence to support any accusation that Best sought IRA "help." In fact, it is not necessary to assume that anyone in the IFAW sought out IRA involvement or knew anything about the ultimate connection. The boycott plan may have sounded so promising as a disruptive force that Soviet interest was aroused spontaneously and the IRA was subsequently directed in this unprecedented involvement.

Lack of complicity is even probable, but this does not ease the news that a major animal rights scheme to influence public economic behaviour has attracted KGB admiration, attention and, perhaps, unsolicited "aid." Animal rights supporters in the free world everywhere should be aware of this possibility when deciding about lending their support to any economic boycotts for a "humane" cause.

Incidentally, since the boycott has died down as a live or effective issue in the United Kingdom, Birdseye Corporation has sent a leaflet to its customers which states that it shall continue to buy Canadian fish instead of obeying the fish boycott mandate and instead of buying from the Soviet Union. The choice, it says, is between buying from a country whose fishermen kill seals and buying from a country which has never respected the human rights of its own or other citizens. This is a courageous statement and a commendable corporate posture. This writer has no idea whether or not Birdseye ever received any IRA cards.

Those corporations in the United States which received advance notice that they would be singled out for the public as targets of the boycott scheme were McDonald's, Burger King, Mrs. Paul's Kitchens, Inc., Taste O'Sea (O'Donnell-Usen Fisheries Corporation), and the Gorton Group. All received the Steve Best fish boycott paper and other documents which allegedly indicated the extent to which Americans could be expected to support the boycott.

In England, those companies targeted had also received "The

LET THEM LIVE

IFAW (CANADA) DONOR FORM

Please fill in your name and address and return this card to IFAW along with your donation in the envelope provided.

Here is my gift now to keep up the fight for animals and to keep IFAW confronting the seal killers.

( ) $1,000.00   ( ) $500.00   ( ) $100.00   ( ) $50.00
( ) $25.00   ( ) $10.00   ( ) Other $_____

Name_____

Address_____

Postal Code_____

2CHNI

Please separate at perforation and keep the picture portion for your records.

SAY GOODBYE

Kiss this baby goodbye.

Non-Profit
U.S. Postage Paid
GREENPEACE USA

Only 12 days to live.

ONLY 12 MORE DAYS TO LIVE
GREENPEACE

YES ...I want to save the seals. Please enroll me as a supporter of GREENPEACE and the seal campaign.

I am enclosing my check for:

☐ $10 ☐ $15 ☐ $25 ☐ $50 ☐ $100 ☐ $500 ☐ Other $_____

I am also sending the enclosed postcards so that the GREENPEACE message will be heard.

As a Greenpeace supporter, you will receive *Greenpeace Examiner*, our quarterly magazine covering Greenpeace's worldwide activities.

All contributions are fully tax-deductible. Please make out your check to "GREENPEACE."

GREENPEACE 1700 Connecticut Ave., NW Washington, D.C. 20070

A copy of our annual report is available upon request from GREENPEACE or the New York State Dept. of State offices of Charities Registration, Albany, NY 12231.

GP-CW4-S

GREENPEACE

YES The Canadian government has your plane, but you have my support. Please enroll me as a supporter of GREENPEACE and the seal campaign.

I am enclosing my check for:

☐ $10 ☐ $15 ☐ $25 ☐ $50 ☐ $100 ☐ $500 ☐ Other $_____

I am also sending the enclosed postcards so that the GREENPEACE message will be heard.

As a Greenpeace supporter, you will receive *Greenpeace Examiner*, our quarterly magazine covering Greenpeace's worldwide activities.

All contributions are fully tax-deductible. Please make out your check to "GREENPEACE."

GREENPEACE 1700 Connecticut Ave., NW Washington, D.C. 20070

A copy of our annual report is available upon request from GREENPEACE or the New York State Dept. of State offices of Charities Registration, Albany, NY 12231.

GP-CW4-A

THE NEW YORK TIMES, SUNDAY, MARCH 20, 1983

100,000 Baby Seals SAVED
through European Import Ban

But Mass Slaughter of older seals is about to begin

The cruel massacre continues! Loveable and intelligent Canadian seals of 4 weeks of age and older will be – **Clubbed – Drowned – Shot,** for money. Anti-cruelty teams from the International Fund for Animal Welfare (IFAW) are now on the ice-floes off the east coast of Canada to fight as hard for these older seals **AS THEY DID TO SAVE THE SEAL PUPS.** But they need your help. Your moral and financial support – **TODAY.** It's vital to the success of our campaign.

The *Boston Globe* said this of the European Parliament's vote to ban all trade in BABY seal products, *"...It's a victory for a concerted campaign by the International Fund for Animal Welfare (IFAW) to mobilize Western Europeans against the hunt."* (Europe is the major market for all seal products.)

But the task of saving the seals is only half done. Please open your heart today. Join a winning team to save the older seals and stop the hunt **forever.**

Please help it's VITAL

I want to save the seals from any more torture. Please accept the donation indicated below to support your anti-seal-hunt campaign.
To IFAW. Here is my contribution in the amount of

$ _____

Name
(please print) _____
Address _____

_____

_____

_____ Zip code _____

IFAW

Please mail your DONATION to the International Fund for Animal Welfare, P.O. Box 193A, (169 Main Street), Yarmouth Port, Massachusetts 02675.
Financial disclosure available on request.   3AUE1

Harp Seal
Prospect: Continued clubbing

HOW THEY WERE MASSACRED

Smashing the helpless infant with clubs until dead is a standard method of killing. Highly-organized huntsmen contended the clubbing would accomplish its aim at a single stroke. Often, it didn't. The baby seals are killed at the age of two or three weeks because the much-prized, snowy-white pelts soon change to a darker color. Since 1940, numbers are down by at least 9.2 million to less than 800,000.

CANADIAN OFFICIALS DESCRIBE THIS SLAUGHTER AS A "CULTURAL HERITAGE."

WHAT YOU CAN DO

U.S. prohibits import of these seal pelts now — other nations should do the same. Before the annual March hunt, lodge your protest with Canada and Norway.

DISCOURAGE FUR-BUYING BY THE PEOPLES OF ALL NATIONS.

LEARNING WHAT YOU CAN DO

The Animal Protection Institute of America keeps a running record of the battle for survival of many world species. It believes the odds in those battles can be changed. For additional information on the threatened and endangered -- and what to do about it — write to The Animal Protection Institute, P.O. Box 22505, 5894 South Land Park Drive, Sacramento, California 95822.

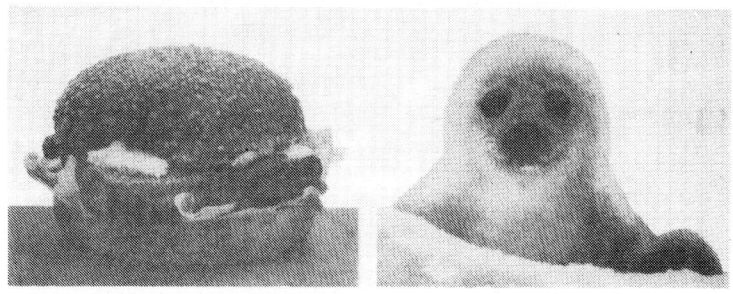

WHY BOYCOTT FISH SANDWICHES TO SAVE SEALS?

If you stop buying fish sandwiches, you can stop the Canadian seal hunt! Why? Because your purchase of a McDonald's or Burger King fish sandwich is helping buy the boats, hard wooden clubs, and guns used by seal hunters as they turn from fishing to the cruelty of killing adult and baby seals.

If you make a pledge today not to buy fish sandwiches from McDonald's or Burger King unless you were assured by the companies that no Canadian fish was used ... your decision would save the seals. Canadian fishermen would have to stop sealing since they are totally dependent on their fish sales, with seal hunting a tiny sideline in comparison.

The Canadian national newspaper, The Globe and Mail, on 13 March 1984 had this to say of IFAW's call for a worldwide boycott of Canadian fish: "...*But what has the officials spooked is the U.S. market. (There) The International Fund for Animal Welfare (IFAW) which spearheads the anti-sealing protest worldwide, has started pressing U.S. purchasers not to take Canadian fish. The U.S. market is worth $1-billion a year to Canadian fishermen and the fast food business in the United States is about a fifth of that market. Canada fears that, if one U.S. fast food chain caves in to the protestors, all will surrender.*"

Over 20,000 seals, mostly pups just 2-4 weeks old, have been clubbed or shot on the ice floes during March and April 1984 ... and yet the Canadian Minister of Fisheries says there is no baby seal hunt!

You have it in your hands to save the seals today.

Please write or telephone one or both of the companies listed below and seek an assurance that they will not purchase Canadian fish until the Canadian Government passes a law banning the seal hunt forever.

Mr. Fred L. Turner
Chairman of the Board
McDonald's Corporation
McDonald's Plaza
Oak Brook, IL 60521
Tel: 312/887-3200

Mr. J. Jeffrey Campbell
Chairman
Burger King Corporation
7360 North Kendall Drive
Miami, FL 33156
Tel: 305/596-7277

IF YOU'RE ON THE WAY IN TO A MCDONALDS'S OR BURGER KING, TELL THE MANAGER WHY YOU AREN'T BUYING A FISH SANDWICH!

These shocking photos were taken by Brian Davies, executive director of IFAW.

IFAW has launched a campaign to boycott Canadian fish products until the torment that seals like these are suffering is ended forever.

IFAW *black death* folder inserted with mailings to public and IFAW membership.

IFAW *black death* folder inserted with mailings to public about the fish boycott.

CANADIAN FISHERMEN KILL BABY SEALS!

THIS IS MY PLEDGE TO BOYCOTT CANADIAN FISH PRODUCTS

Photo: Daily Mirror

In order for the message to be effective with housewives, it would have to be "dog simple."

HELP THEM LIVE

BOYCOTT CANADIAN FISH PRODUCTS

To: INTERNATIONAL FUND FOR ANIMAL WELFARE
PO Box 193
Yarmouth Port, Massachusetts 02675 USA

I pledge my support for the boycott of Canadian Fish Products. Here is my gift now to save the seals...and to help IFAW continue its fight for all the seals, whales and other animals.

( ) $1,000   ( ) $500   ( ) $100   ( ) $50
( ) $25   ( ) $10   ( ) OTHER $_____

Your contribution is tax deductible.

IFAW

12801

Please separate at perforation. Keep and display the picture portion as a reminder of your pledge to boycott Canadian fish. Please return the labeled form along with your donation to IFAW in the envelope provided.

Stop It!

I'll do my part to help stop the atrocious slaughter of baby seals. I've signed and mailed my postcard. Now, I want to help the International Fund for Animal Welfare launch a hard-hitting *Canadian Fish Boycott Campaign*.

Enclosed is my tax-deductible contribution in the amount of:

☐ $10 ☐ $15 (☐ $25) ☐ $35 ☐ $50 ☐ $100 ☐ $500 ☐ Other $_____

WE NEED HUNDREDS OF GIFTS THIS SIZE TO HELP SAVE THE BABY SEALS — BD.

Dear Mr. Turner,
  Canadian fishermen are still killing baby seals . . . I will not buy your fish sandwiches until you assure me they are not made from Canadian fish.

Signed _____

Address _____

_____

_____

Mr. Fred L. Turner
Chairman of the Board
McDonald's Corporation
McDonald's Plaza
Oak Brook, Illinois 60521

EEC TO BAN SEAL IMPORTS IF BRITAIN AGREES

A letter to your MP is vital!

An appeal from Brian Davies, Executive Director, International Fund for Animal Welfare.

Dear Reader,

Although the clubbing of the baby seals is over for this year, older seals now become the victims of the mass, merciless slaughter – 201,000 in all!

They are shot – and I have personally witnessed the very ocean stained red with the blood of terribly wounded animals.

They are caught with nets and traps to drown in the icy depths – surely, for a seal, a horrible way to die.

All this cruelty primarily to provide fur and leather for the European market.

To help end this suffering it is vital that you write to your M.P., today.

Thanks to letters from animal lovers and conservationists throughout Europe (30,000 came from Britain), the European Parliament has recommended that the EEC ban the import of seal products.

Post this request TODAY to your M.P., House of Commons, London SW1. Ask your friends to write to their M.P.'s as well, or photostat the statement below for their use. As more people write, more M.P.'s will give their support!

This proposed ban, however, must have the approval of the British Government and of other EEC countries before it becomes law. Important motions to protect the seals are now before the British Parliament. They need the support of every M.P.

For the sake of the seals please write a letter to your M.P. asking him or her to sign Early Day Motions Nos. 342 and 344. This is the best help you can give the seals today. But if you haven't time to do this, then I beg you to sign the statement below.

The EEC imports 75% of the products of the Canadian Seal Hunt. A ban on these imports would turn the hunt without pity into the hunt without profit – and end the commercial slaughter forever. PLEASE TAKE ACTION TODAY.

Yours sincerely

P.S. M.P.'s are busy people. Tell your M.P. a reply is not necessary if he or she has signed the Early Day Motions.

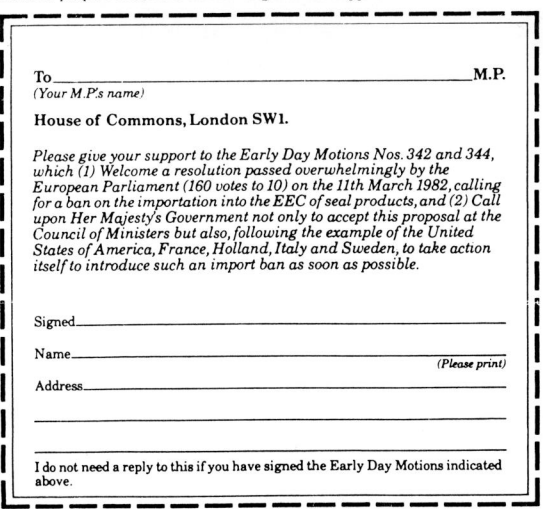

To _____ M.P.
(Your M.P.'s name)
House of Commons, London SW1.

Please give your support to the Early Day Motions Nos. 342 and 344, which (1) Welcome a resolution passed overwhelmingly by the European Parliament (160 votes to 10) on the 11th March 1982, calling for a ban on the importation into the EEC of seal products, and (2) Call upon Her Majesty's Government not only to accept this proposal at the Council of Ministers but also, following the example of the United States of America, France, Holland, Italy and Sweden, to take action itself to introduce such an import ban as soon as possible.

Signed _____
Name _____ *(Please print)*
Address _____

I do not need a reply to this if you have signed the Early Day Motions indicated above.

IFAW

I enclose a donation of   2ENA5
£_____ to help you finance this crucial campaign to stop the seal hunt.

Name _____ *(Please print)*
Address _____

Please send this coupon with your donation to:
International Fund for Animal Welfare,
Section 62, Tubwell House, New Road,
CROWBOROUGH, East Sussex TN6 2QH.

Do not enclose this coupon with your letter to your M.P.

Gallup Omnibus Report," a study conducted in September of 1982 in which Gallup Poll staff asked Britons about their fish eating habits and their attitudes on the killing of harp seal pups by Canadian fishermen. In the United States, targeted companies received a September 1982 research report prepared by Nationwide Market Research Corporation entitled "Attitudes towards the Harpseal Hunt." It performed the same function as the British study and served as a further scare tactic to discourage American importers of Canadian fish from renewing their orders. Steve Best sent each to this writer, probably in order to impress me with the likelihood that the fish boycott would be a great success.

In January of 1984 the IFAW claimed that boycott information was mailed to 7.5 million Americans. Targeted corporations began to receive pre-printed postcards protesting the seal hunt and the use in the United States of Canadian fish. Predictably, some corporations asked the Canadian Department of Fisheries and Oceans what it would do to alleviate the pressure. They wanted to know if the hunt would be called off so that the public would forget the whole thing.

Canada's east coast was in a turmoil, not knowing what to expect from Americans who had always swallowed the protest account of sealing without question. There was some discussion of actually cancelling the seal hunt in order that the entire thing be defused. It turned out that such a move was unnecessary.

The Federal Department of Fisheries and Oceans and the Canadian Sealers' Association released announcements that white seal pups were no longer being clubbed, in response to demand. Although the "demand" was a result of the loss of an European market, Americans didn't understand that whitecoats were not a commercial item anymore and those few who had heard about the boycott assumed that the killing had stopped in response to it.

The *New York Times* picked this up on March 9, 1984 and ran it as a double-topic story about the hunt and the boycott attempt in the United States. It announced that whitecoat pups were no longer being killed and that McDonald's headquarters in Chicago had announced that it would continue to buy Canadian fish. The public read the story and believed that the hunt had been officially cancelled by Canada.

It was all over now. Brian Davies' claim that "Canadian Fishermen Kill Baby Seals" had been effectively defused and the public believed that the boycott was responsible for the announcement. Of course, it was. The public wanted to believe that the boycott had been effective.

They didn't realize that they had been had both ways. The clubbing of whitecoats had effectively ended with the 1982 season, but Davies had been telling them otherwise. And Canada and the Canadian Sealers' Association had decided that even though a few young pups were still being taken "for subsistence" and for the hair seal trade, they could afford to announce that the whitecoat hunt was over in order to save fish sales. This may have been a brilliant defensive move.

In the lower Hudson Valley, a typical response to the recent media coverage was seen at a local humane society meeting; the leader stood up and made the dramatic announcement that the baby seal hunt was now over. Applause was enthusiastic.

A second IFAW fish boycott mailing was sent to Americans, however, and it was claimed that Americans should continue to check the labels on their grocery store fish products in order to avoid Canadian sources of fish. The second IFAW packet, received by 7.5 million homes, called for a boycott of the five targeted companies and the sending of letters of protest until the hunt was officially cancelled.

An ad in the *New York Times* asked if consumers really knew what went into their McDonald's fish sandwich. Pictured was a harp seal pup and a fish-in-a-bun. The message stated that seals were still being killed by Canadian fishermen and that purchase of this Canadian fish product actually encouraged Canadian fishermen in the killing of more seals. Thus, Americans should refuse to buy the McDonald's product.

On March 28, 1984 a small piece in the *Wall Street Journal* outlined the situation:

FISH FIGHTS
McDonald's Corp. is undeniably responsible for the deaths of millions of cows, chickens and fish each year. But baby seals?

The article went on to explain that an IFAW spokesman justified the picketing of McDonald's in a few cities because "Canadian fishermen are the ones who are killing baby Harp seals."

The IFAW reportedly planned to mail pictures of clubbed seal pups to those who lived near McDonald's outlets in hopes of generating public pressure against the company. It was assumed that McDonald's would then pressure Canada to cancel the hunt.

Richard Starmann, a McDonald's vice-president in charge of purchasing, claimed that officials "had been assured by the Canadian consul in Chicago that the hunt was being conducted properly." McDonald's would not cut back on its Canadian fish orders in spite of an expected "onslaught" from seal supporters.

It is not known if the corporation began litigation procedures in response to the IFAW "tainted fish sandwich" ad, the effect of which had been shocking and bizarre. Reportedly, an IFAW attempt to buy space for it in a Florida newspaper had been refused on the grounds that it was in poor taste, if not actually libelous in both intent and content.

By mid-summer there was still no news of orders for Canadian fish having been cancelled by any American buyers in response to IFAW blackmail threats. In spite of the tremendous amount of advance hype which had been sent out to the five targeted corporations, they had all elected to wait and see how the public would react before complying with the mandate.

In retrospect, perhaps the entire thing had been an elaborate bluff, a power play so seemingly enormous that it might have worked without the IFAW actually "firing a shot" in the form of a really well financed paid aid campaign in either the UK or the US. In support of this theory is the fact that the mailings were so small; a total population of some 256 million people had been virtually ignored by a direct mail campaign which was aimed at only thirteen million households. Furthermore, hardly any paid advertising was seen in American newspapers. The "seal sandwich" ad was so preposterous that it was almost laughable and may have had some value as an indication of things to come but had no effect on the public's consumption of fish sandwiches at McDonald's.

Davies must have known that typical response (card sending, letter writing, or donation rate) was always very small and that the .2 of one percent usually elicited would not actually have hurt fish sales. The plan may have been to generate fear and belief on the part of the Canadian government and fish wholesalers and retailers alike that widespread public support was probable. Fear within the Canadian fishing industry itself could have led to demands that the government cancel future seal hunting before orders could be lost.

This fear was probably heightened, knowing that Davies had put three million IFAW dollars into major newspaper advertising during the European pelt ban effort and might do the same in the United States.

Word was also circulated in Canada that a bill in the United States House of Representatives was being promoted by Congressman Jeffords of Vermont (Jeffords had previously supported the "Save The Seals" campaign of the Humane Society of the United States). His House Bill 5032 proposed to make mandatory the labelling of all imported fish products by country of origin. This would have enabled

American housewives to easily identify, and reject, Canadian fish. The bill went nowhere, but it gave Canadians cause for worry.

Monday morning quarterbacking of a football game is not a risky undertaking because the game is over. Major American fish importers will probably not engage in quarterbacking on the boycott issue because there is always the possibility that the game could begin again; the IFAW could always "rear its ugly head" with a massive advertising campaign so that the American people would be more apt to pay attention to the boycott.

Was it all a bluff? Only the IFAW knows for sure. But some things are certain; the Canadian Fish Products Boycott turned out to be at least a temporary image booster for the IFAW. It caused great consternation and fear among fishermen on both coasts and threatened to be divisive as a force which overrode their common interests.

It influenced corporate decision-making in Great Britain and the United States and at least delayed some Canadian fish orders for awhile. It caused innumerable high level meetings and argument between the Department of Fisheries and Oceans and the Department of External Affairs and was the subject of countless thousands of internal and international telephone and telex messages.

The apparent reluctance of External Affairs to publicly discuss the effect of the boycott threat was a source of great dissatisfaction at home among fishermen whose livelihood was definitely threatened. Of course, all fishermen believed that the government had done nothing to avoid or stem the problems caused by the boycott. Feelings that the government had let them down pervaded all talk among fishermen when the topic of protest against the seal fishery was raised.

The fact that the price support board had decided to give sealers a payment to make up for eighty percent of their losses on pelt sales in 1983 was considered temporary relief, a bandaid token in face of everything else which appeared to be wrong with government policy on defending the hunt.

The boycott had kept the seal issue before the public, even though media coverage was almost absent due to loss of the European market for pelts. The boycott mailings generated some further income for the IFAW, and kept alive its image as a humane-oriented organization which was internationally powerful. It was still an economic and political force to be reckoned with.

The IFAW boycott plan had again convinced many people that their humane wishes would be heard, even in the face of big business and national government policy. The fact that the IFAW briefly convinced so many that the cause was worthwhile, or that it would work regardless of real justification, was perhaps the greatest victory. Certainly, its own 500,000 members would not abandon it just because the cruel Canadians might still go sealing. It had risked nothing in the sense that member loyalty would never fade. It did not matter whether its boycott scheme had been serious, or a bluff. Credibility would remain high, because to IFAW members it was still obvious that Brian Davies' intent was to save the seals.

Davies had made history by generating public pressure which influenced a major legislative body. Removal of market opportunity no longer depended on influencing consumers not to buy the product. Now the product itself was outlawed by public demand, at least temporarily. Furthermore, it was not the stated intent of this demand to have a biological impact on the species as a whole. The justification was the prevention of cruelty to immature animals and their mothers.

The implications of this for the future of wildlife management of fur bearers are staggering. Scientific assessment of animal populations can never mean as much to the general public as repetition of claims that the animals are taken in a cruel manner. Not only the manner of taking, but the fact of taking becomes an issue easily promoted by an anti-fur protest movement.

Public funding for wildlife biology, habitat research for any game species or species control of any kind shall always be influenced in the future by the threat of involvement of an "animal rights" protest organization. It has been demonstrated that such groups have substantial, perhaps perpetually renewable, financial resources.

Davies' precedents in influencing the United States Marine Mammal Protection Act, and the European pelt ban, should not be ignored. His type of influence will certainly not go away as long as there are people to donate to any "humane" cause which is presented to them.

In 1958 William J. Lederer and Eugene Burdick wrote *The Ugly American*. This novel is about the clash of values which resulted when Americans began to move into the Orient after World War II, and to exert their influence in both the economic and military spheres. The insensitivity of all nationalities of western men insulted, hurt, and otherwise thoroughly offended nearly all those people who were contacted. The reason was an inability on the part of "Ugly Americans" to perceive anything of value in the less dominant cultures. Ugly Americans, in this sense, typically attempt to force their own values and beliefs on all others, and to deny that other cultures have the right to continue pursuing their "substandard," backward traditions and customs.

The role which Brian Davies has played in creating an International Ugly American should never be forgotten. He has convinced millions that the power to intervene in another culture's way of life is entirely justified by one's own standards of humane treatment and decency.

This philosophy of the unlimited right to intervene, even at great human cost, is of monumental significance. It contains no room for cross-cultural understanding, no possibility of compromise, and no respect for each culture's uniqueness. The Ugly American, of any nationality, is sure that economic and political power *per se* justify the forced cultural change which he or she desires.

The International Fund For Animal Welfare may be contacted at: Post Office Box 193, Yarmouth Port, Massachusetts 01675, U.S.A.

In 1984 those who inquired about the status of the fish boycott received sample postcards, "mini posters," large, attractive buttons, and petition blanks on which to collect signatures and addresses to be sent to the McDonald's corporation.

School children who wrote to IFAW were advised in detail how to proceed in organizing a picket line in front of their local McDonald's. The IFAW stated that if a time and place for a demonstration was given it in advance, then it would notify the local newspapers, radio and television stations. Large posters would be supplied, which should be mounted on stiff cardboard and could be displayed in front of local McDonald's outlets.

It is obvious that the IFAW realizes the value of cultivating humane interest in all children. They will make excellent Ugly Americans some day.

VII

Victims of the Seal Wars

The following letters are presented exactly as written, except that the name of the Minister of Fisheries and Oceans has been omitted. Some letters were turned over to federal security since they were indicative of an ugly intent on the part of the writer. Some, of course, were from ordinary citizens who were concerned about the seal slaughter which they had seen on television. Some were from children who had learned to protest the hunt from either television programmes or from their teachers, who in many cases have been their role models in protest. All are unedited, real messages from people who felt a genuine rage or concern. These are some of the victims of the seal wars.

Dear Minister,

I am writing to you about the killing of the baby harp seals. I think this is an inhumane act. Please stop these killings.

The baby seals are young and cannot defend themselves. These animals have the right to live as any animal or human should.

Instead of killing these animals you could take pictures or make posters of them to sell.

Thank you,
Pam
[Island Lake, Illinois]

To Whom It May Concern:

You better stop killing the seals! It's very cruel. If you don't stop killing them they will become extinct. If that's what the people over there think about making money by hitting baby seals on the head with a club it won't work much longer; people are showing what they do to those seals everywhere — on TV, on the radio, etc. & they're not buying things and stores aren't buying things either. Those seal killers are some of the worst people in the world. So, why don't you just stop killing seals?

Amy
[Ohio]

Greetings!

Last night of 19 March 1981 ABC NEWS NIGHTLINE showed and dealt with the horrible and sickening baby harp seal slaughter that takes place, I believe, every winter season under the auspices and protection of the Canadian Government and the Department of Fisheries of Canada.

Seeing the Canadian and Norwegian Yahoos and goons wantonly killing the baby harp seals for filthy lucre and greed is both disgusting and revolting...and to think that all this is abetted and aided by the so called civilized Government of Canada. You all need moral introspection!

For your better consciences,
Albert
[McKeesport, Pennsylvania]

Dear Minister:

In the spirit of good neighborliness, and in the spirit of justice and mercy for helpless creatures, I earnestly and sincerely again implore you to stop the tortuous clubbings of baby harp seals.

We Americans have always regarded your people as compassionate and just; please take another soul-searching perspective of the brutal clubbings which go on year after year. Surely, there must be a way to provide a livelihood for some of your citizenry other than the field of savagery.

May this protest awaken consciousness to the side of righteousness, and SAVE THE SEALS.

Sincerely,
Mrs. K.
[Mt. Prospect, Illinois]

Dear Sir,

I can't let another seal hunt go by, without registering my protest against this needless slaughter. It just seems to be incredible that any human being could be so insensitive as to club a baby animal to death. Our Canada with its plentiful natural resources should be in the forefront to conserve such a beauty.

Yours truly,
Wendy
[Parry Sound, Ontario]

Mr. Minister:

Fuck you, you loathesome ass hole. Let me assure you that one of these fine days someone may just bash your brains out in the same way you and your cheapassed government permit the seals to be massacred. Let me add that it won't be by me, because I may have stepped in some American dog shit and I have too much respect for the shit to bring it into contact with the soil of your disgusting country. I am not only exerting every effort against your scummy fish, but against every Canadian product I can find. Activist takes on a new meaning when describing my hate for you and your dime store dump of a country.

Hatefully and vigorously,
Spencer
[Hollywood, California]

Dear Sir:

You should be ashamed to allow fishing companies to make their profits out of cruelty to animals, leading every year to the wasteful and terribly inhumane Baby seal hunt; Yes, I have seen movies, they start peeling them off still alive! They also net pregnant mothers as they swim ashore to give birth. All those who agree or participate deserve to suffer some day as much as they have made those poor victims suffer. And I hope they do!

Nelly
[Menlo Park, California]

Dear Sir,

I was horrified to see on Australian 60 Minutes program about the slaughter of baby seals which are defenceless and hurt nobody.

I wish to lodge my protest with your government against the slaughter of the baby seals and plead with you to stop it because it is inhumane.

Any self respecting government would stop the slaughter. Let those greedy people go without their fur coats. Trusting you reconsider your laws in this matter.

Yours faithfully,
Harry
[Urunga, NSW, Australia]

[The following is a translation of a letter from children in Hoogvliet, The Netherlands.]

Dear Members of the Government of Canada,

I want to say something about the seals.

I don't think it is necessary to let the seals be beat to death by the crush-hats.

There is no need to. And I think you must put an end to it. Otherwise there will not be one single dear baby seal left, they cannot live then. Don't you think likewise?

That is why I — together with all my schoolfriends — hope that you will let them live happily.

Life must be nice for men and animals: you want to live long and that is what the baby seals, too.

And I think that the young seals are treated painfully and meanly.

Written by a nine year old kid.

[Signed by some eighty school children, and apparently sent by an adult on their behalf. Translated in Canada.]

Dear Minister:

When will you see to it that Canada joins the ranks of civilized nations. You are the greatest environmental criminal in the western world. Stop slaughtering seals, you bastard.

Sincerely yours,
Howard
[Chicago]

Dear Sir,

My name is Diana. I go to Harrison Public School. I would like to say "Stop The Seal Hunt." I saw on Those Amazing Animals how the people kill harp seals. I know that is a part of their living. But if you were a woman and you had a baby you would not like it if someone came along and pounded your baby over the head only for its skin. "Would you like it?" "No." So I and hundreds of people would like to see the harp seal hunt end. Harp seals are not only harmless animals they are number one on my list of beautiful animals.

Yours truly,
Diana
[Georgetown, Ontario]

Honourable Minister:

Please, please, stop the killing of these seals. It is a criminal act.

The great dom. of Canada is surely above this kind of slaughter. Your esteem, world wide, has greatly diminished.

Again, we are very shocked at this wanton killing of the seals.

Very truly yours,
Robert
[Quincy, Massachusetts]

Re: Television ABC program on 3/19/81, 11:30 p.m.

To Whom It May Concern:

I joined the Animal Protection Institute of America and since then I have been thinking about the animals' rights. They seem to have a percentage of zero!!

I have always been an all animal lover and went into tears when I heard and read that the baby seals were being clubbed to death and even skinned alive for a stupid thing called money! My mother has always said that money is a pathway to evil. I must say that I agree to the fullest extent!

I saw a film last year in school about wildlife and their pathway to extinction. I have reason to believe that the end of all wildlife will come before my life is over. (I am thirteen years old.) UNLESS YOU PEOPLE DO SOMETHING BESIDES WHAT YOU ARE DOING NOW, THE PREVIOUS STATEMENT COULD BE TRUE! LET'S NOT DESTROY SOME OF THE MOST BEAUTIFUL THINGS IN THE WORLD — WILDLIFE!!!

A very concerned citizen
Janet
[Lakewood, Colorado]

Dear Minister:

I am writing in protest of the seal slaughters. This kind of occupation is ludicrous in the world of today — it is unnecessary, stupid and dumb, and inhumane on top of it all. These slaughters should be and can be stopped and end so terribly much grief in the world for so many seals and people.

Sincerely,
Mrs. Carol G.
[Washington, D.C.]

Dear Minister of Fisheries & Oceans,

I am saddened by the fact that I have such a wretched reason to be writing this letter. However, it is a necessity as far as I am concerned to which attention must be given. I am speaking about the clubbing of seals.

Why has such evil activity been allowed to happen in your country? Where has the universal concern for life gone? How has it been able to be replaced by such selfish greed? It may be that my love of all living beings makes me unable to see the viewpoint of the man (?) who, with his club, beats to death a seal for whom he has no grudge. He is a man (?) who in turn with his paycheck supports his narrowly selfish material lifestyle.

I realize that the tone of this letter is judgemental. I understand that I am not in a position to TRULY judge the righteousness or evilness of this issue. I do however claim a right to voice my very strong opposition to the clubbing of these innocent creatures of God.

Most sincerely concerned,
Meredith
[Winston-Salem, North Carolina]

Dear Sir,

I read in the Montreal Gazette that those of us who oppose the slaughter of baby seals are blackmailers. I am an unwashed — so-called — blue collar worker who has never blackmailed anyone in his life. Should you care at any time to repeat this statement up a back alley I will be more than pleased to show you what getting slugged is all about. Maybe you will then have some idea what the poor little bastards are subjected to to provide 'fun furs' for the idle rich.

Yours sincerely,
John
[Montreal]

Dear Stinking Liar,

My friends and I read your blasting attack on the animal protection groups, particularly Brian Davies and we must say that you have strong nerves to act as you would be a big shot.

Since you blackmailed the Europeans about fishing rights you have no bloody right to talk about blackmail yourself. It is a fact that your foul government sticks with business and big industry to pull as much profit as possible from the seal hunt while disregarding the matter of environmental protection.

In fact you are a foul smelling liar yourself. You tell the Europeans that the Canadians want the seal hunt. Well you should fuck yourself for these lies. Please organize a gallup poll and see what Canadians want. 60% do not wish the seal hunt to be continued. So you dirty bastard that god should send to hell including your children should listen to public opinion instead to dollars and industry who care a damn about the environment. I hope you will break your neck soon. You are a pain the ass, that's for sure. Shut your mouth and don't talk about things you don't understand.

Somebody who cares about a healthy world,
Frank
[Lachine, Quebec]

Dear Sir:

I find it hard to believe that this country is still allowing the bludgeoning of baby harp seals on its coastal waters. The practice is disgusting and inhumane, and I am ashamed in terms of the killing of seal pups, to be a Canadian.

It serves NO PURPOSE to kill them except to cater to the whims of people who want white fur! Phooey to them! The Canadian government should not turn its back on defenseless seal pups, or soon there will be no more seals to defend! If those people?!? continue to slaughter the offspring of a species, the species will die. Please don't let this happen! Don't let the GREED for white seal pup fur destroy a species or perhaps disrupt the balance of nature. The nation's wealth is not increased by this slaughter, instead my national pride is insulted because of this gross injustice allowed by the government of my country.

Please consider people's feelings and help stop the seal slaughter. Thank you for listening.

Sincerely,
Mrs. Karen S.
[St. Catherines, Ontario]

Dear Sir:

I am against the killing baby harp seals. I wish you would stop killing the baby harp seals. I think that when they skin them when they are still alive that it is the cruelest thing I ever heard. And it also causes its mother to be sad, because when they come out of hiding, she sees the bloody mass and knows its her baby and holars like you wouldn't believe. Baby harp seals are harmless. They don't even got a chance to live. So please, stop killing the baby seals off. Thank you.

Yours truly,
Tammy Ann
[Watsontown, Pennsylvania]

Dear Sir:

I and my friends think that the seal hunting should stop. We think that seals should have a right to live. Seals are a beautiful thing. and we kill to much. I have never seen a baby seal in my life and I havent seen a big seal either and if you keep on killing I may never see a seal. Please stop killing seals.

Thank you,
Stephen
[Lake Cowichan, British Columbia]

Dear Sir:

On reading the article in The Herald, our daily newspaper on the subject of baby seals, I am amazed that you and your fellow countrymen are able to sleep at night with this on your conscience. I realize people are in business to make money, but in this day and age where most products can be made from synthetics, is this mass slaughter really necessary? I also realize that over population of a species can not only create problems for itself but also for other forms of life as well.

Surely there must be some way to ensure that if they must be killed, then it can be done with minimum suffering to the animal concerned. There must be some way to ensure that all responsible for such killing have concern for the animal involved regardless of the amount of money it will put into their pocket.

On reading the report of seals being skinned alive, I was disgusted to think that any person with any claim to sanity could be a party to such a practice. I am sure that I am voicing the opinion of every person who read this report, when I say that those persons who have the authority to do so, must take the action needed to make sure that this deplorable method of killing is eliminated.

Yours faithfully,
Mrs. G.J.L.
[Balaclava, Victoria, Australia]

Dear Sirs,

I am a grade 4 student and this is my first letter to the government. I am writing about the terrible killings of baby seals that I just saw on television on a show called, "Those Amazing Animals" and I just want to let you know how much it upset me, that I would wish that you would make them stop this cruelty because all they are getting are little souveniers and things we don't need. Please make them stop this cruelty. Well thats all I have to say.

Yours sincerely,
Angie
[Calgary, Alberta]

The following are letters to the Canadian Ambassador to the United States.

Dear Sir:

I would like to take a moment of your time to express my horror at the ongoing practice, in your country, of murdering seals. This act is representative of the very thing we strive to teach our children against — animal cruelty, in the purest form.

If people were starving and seals were the only available food source, a humane killing could be tolerated. This is NOT the case. I agree seal skin is beautiful, but would it not be more beautiful left on the animal it was created for? Has money become more important than life? God help us if this is true!!!

I am only one person, however, I can, and will help spread the word telling people of your cruelty, inhumane treatment, and downright greed. If our only immediate tool against this slaughter is a boycott of Canadian products, then I shall whole-heartedly join in that boycott. I shall also work to gain increasing support for this boycott.

Please reconsider allowing this abhorrent treatment of life to continue!!!!!

Thank you for your time.

Susan
[Dallas, Texas]

Dear Whomever,

I am hereby boycotting all fish imported by Canada. I feel very strongly against the brutalization of these white seals. Its cruel and not human to do this to creatures that can't fight back. And the way you murder these cuddley cute creatures is unreal! Put your fishermen to some other use! That's all I can do but I hope it helps the seals.

From someone who hates you!!! and what your doing!

Seal lovers,
Crystal
Reney
[San Jose, California]

Dear Ambassador,

I am in the seventh grade at Britton Middle School. I am very ashamed that people are and can have the heart to beat seals on the head until they just die. I don't think you would like it very much if someone came along and beat you on the head with a bat just for your hair or skin. I almost cry when I think about it. I am going to boycott all Canadian fish products until the killing is STOPPED!!

Thank you for reading this letter.

Amber
[Morgan Hill, California]

Since 1965, hundreds of thousands of people have been deeply affected by the seal wars. As with most wars, the sporadic battles have been fought in seasonal offensives which became increasingly destructive, and more wide-spread. And as usual in war, women and children have suffered more than their share. It may be said that they have been as much victimized by those who carried the fight as those against whom the offensives were launched.

Many of those women and children, ironically, were in the front lines without knowing how much the outcome would depend on their performance. When told by the various protest organizations that their concern and participation would help to save the seals, they gladly gave of themselves, heart and mind, without question, and of their money, until the pennies and dollars became a flood.

That flood, cresting each spring, found its wellspring in the environmental movement of the '60s. Suddenly there was wide support for the "environmentalists" who pointed with growing concern to signs that the earth was being spoiled beyond imagination, beyond repair, by the unprecedented and unstoppable growth in numbers of the one species with the power to destroy the planet.

Suddenly the pendulum slowed from a joyous, unchecked pride in progress, down-arc to the beginnings of a deep-seated horror of using up the earth's resources. Neglect, overexploitation through ignorance, poisonous waste dumping, and plain unmitigated greed began to be recognized. The accelerated extinction rate of both plant and animal species was seen as proof that the earth needed to be cared for, protected from man, before the qualify of life, and indeed, life itself, should be lost. The new up-swing gained strength and momentum.

Those skeptics who, through the years, have described most environmentalist fears and beliefs as all wet, have finally realized a few arrid facts: wildlife management and other forms of conservation, though proven effective, often have not been perceived by the public as making any difference to a threatened natural world. Part of the reason for this is that most of the information given to the public comes from the humane- protest/environmentalist movement. Professional wildlife management has failed to command respect in whole populations which have no experience with it, and so no tradition of belief in the field.

Those biologists who have been stewards of the earth's renewable resources have seen their credibility publically denounced and ridiculed by the protest movement. This could not have happened at a worse time, for their pooled knowledge and record keeping have finally begun to show that marine resources can be skillfully manipulated to benefit not only man, but the entire environment.

The public has been told that the seals and whales will become extinct if something is not done to stop their exploitation by man. It is not in the interest of protest to ever admit that professional scientists know what they are doing. It is in their interest to talk about numbers of seals before commercial hunting and to insist that they have dropped by fifty percent since the 1800s. The fact that the fish biomass has also decreased as man learned to take it more efficiently is not fully addressed, except in the context of blame for greedy "overfishing."

The sudden environmental impact of a ban on the hunting of whales and seals is undeniably present in a growing pressure on the marine ecosystem, yet this is not discussed as important, even by those who tout "balance" as a primary goal in the environment. Those seals and whales which become entangled with increasing frequency while raiding fishnets are not described as overabundant, unfortunate nuisances, but as victims of greedy and uncaring fishermen.

In all protest literature the needs of those people who live closest to the marine resource are minimized, even to the point of claims that fishermen need nothing from their shores but fish. Their right to take seals and whales for whatever reason is denied entirely by a protest movement which has enforced its claims with powerful legislation, produced with public pressure wherever it would have the desired effect.

Over the past twenty years millions of anguished and angry letters from around the world have flooded the offices of Canadian embassies, newspapers, parliamentarians, and the prime minister. Thousands have landed in the offices of the Department of Fisheries and Oceans, where they have been duly answered, noted, counted, and stored. These have ranged from polite requests that the seal hunt be stopped, to obscene, threatening demands, to desperate pleas from children. The educational range of those who write is extreme, the themes are narrowly universal: the killing is cruel, unnecessary, and apt to cause harp seal extinction.

It may fairly be said that good people, feeling some guilt for the powerful impact of their burgeoning civilization, have transferred their sins to those who deserve blame the least. Urban man has been the farthest removed from nature, and has romanticised it the most. His acquaintance with animals has been intimate, however, in the traditional patterns of pet ownership. We deeply love our dogs, cats, birds and horses. We believe in the emotional capabilities and intelligent reactions of our domesticated species. Because of this, most of us have lost the ability to recognize that all creatures are endowed differently due to the differences in their evolutionary course and environmental influences. Our myths about superior intelligence and "sensitivity" in seals, whales, sea lions and dolphins have all sprung from that blurred visual handicap which is the heritage of blind love.

Since the environmental movement began, thousands have gone out in small boats to hug and pet migrating gray whales. This has been an emotional experience, an unprecedented, unforgettable outreaching of one species to another. Some have played flutes to whales, which have responded with their own vocalizations. The whales' specific reaction, although not understood, is often considered to be love and a desire for communication. The possibility that whales show mere bovine-like interest in and tolerance of an apparently harmless, increasingly frequent species is not very satisfying to adherents of the myth.

Many are unaware that genetic programming in wildlife enhances recognition of both food species and enemies. Since whales evolved in environmental conditions isolated from homo sapiens, that programming shorts out on us; we fit neither category. With land mammals it is entirely different: try starting a "hug the grizzly" movement and see how far it goes!

Adult seals usually flee from any quadruped, or from any creature which moves across the ice. Put a human in the water, however, and the programming fails to function. The evolutionary history of the seal prevents it from recognizing a danger from a swimming mammal its own size. The whale, likewise, is not programmed to recognize a swimming human as either food or threat.

Self-preservation is a part of seal programming and their sensors tell them when we are in a predator category. Harp and hooded seals evolved in a time before man could exploit them out on the ice. Few other predators of note made it out onto the whelping grounds. There was no survival advantage for any pups to exhibit a flight response, to be programmed to flee a mammal. Thus, seal pups fail to react to humans in any way which looks like fear.

This behaviour has been labelled through the eyes of a removed civilization as "friendliness," "sensitivity," "innocence," and God-given trust in man. As soon as seals mature, however, their synapses line up appropriately to serve them in a new environment. There, predators of many kinds may be expected to interrupt survival. Try to corner and hug a beater, bedlamer, old harp or hooded seal; s/he will eat your lunch. This is not behaviour learned in response to the animal's life adventures. This is survival instinct.

Thousands of good people have been victimized by the seal wars in the sense that they have been given less than adequate and sometimes completely incorrect information with which to make a decision on how to feel in this matter. They have truly believed what the animal rights movement has taught: that cruelty, waste, and species depletion are facts which demand a total permanent end to all use of all marine mammal resources. The "Save the Seals!" slogan carries the promise that at least one of the sins of the world can be righted.

This message has been growing in strength and intensity since 1965.

Twenty years after the protest against seal hunting started, the voices of the other side began to be heard. These other war victims are the seal hunters: Alaskan, Canadian and Greenland Inuit, and eastern Canadian fishermen. Like most people, they have lived their lives without worrying about the future until a threat to it was perceived.

The new pattern developed slowly; first it was recognized that outsiders were, for some reason, complaining that seals were being killed in a cruel manner. The concept of cruelty to animals was entirely unknown to the Inuit whose tradition it is to use whatever resources are available and to deeply respect them. Animals are all sent by the spirit world for humans to use.

Newfoundlanders themselves found it hard to believe that outsiders could complain about the way seals were killed. After all, it was very close to the way hogs and beef were slaughtered, and surely the rest of the world used those animals. It was felt that this was strange and unimportant.

Because all those who hunted seals needed to do so for direct subsistence and for cash flow reasons, it was a long time before any imagined that such an important part of life could, indeed probably would, be taken away from them by those who failed to understand.

The world knows that the Inuit live in special conditions. Yet there is still resistance to the idea that they should be able to make money from ivory carvings, sealskin items, whale or polar bear products. The trend of white civilization has been to force the Inuit to abandon their dispersed settlement pattern and come together in central communities where it is more convenient to give them medical care, welfare checks, and use them in occasional manual labour jobs in mining and oil exploration.

Inuit children are taken from their homes and forced to learn in the white man's educational system. Hunters find that in order to get out to where the food resources are they must have snowmobiles. Modern artifacts such as skidoos, rifles, outboard motors, television sets, juke boxes, alcohol and tobacco all have to be bought with cash. Until 1983 cash came from seal skins. Up until this time, while life was not ideal by any means, it was possible to live fairly well and to enjoy the material comforts available. Because there was cash to get out to the resource area one could still hunt the bear, caribou, whale and seal. Native food was better than any other, rich in the B vitamins, iron and other minerals.

Boys learned to hunt and the virtues which attend that necessary part of life; girls learned to make the skin clothing. There was no confusion in sex roles. Modern medicine increased the life span and decreased infant mortality. A person could now look forward to living long beyond the thirty-five years which was the average only a generation ago. There had been a start at melding the elements of old and new cultures so that a workable accommodation to the ways of the Kabloona (who wouldn't go away) was becoming a reality.

The loss of the market for sealskin changed all this. Even though the ban in Europe was on products which the Inuit never handle, the ripple effect extended to all sealskin and the bottom fell out of the system.

A person has to have money in order to buy the gasoline, ammunition, and machinery which is now necessary (because the family lives in town) in order to get out to hunt. If there is no money to be had from sealskin, then a family has to live on the welfare check, and it is not enough.

A man cannot hunt. Therefore, there is no seal meat, the best food in the world. There is no material for new clothing in which to be proud. There is nothing to teach one's sons or daughters. There is only money enough to buy store bread, store bologna, pizza, and beer. Everyone who cannot get out to hunt finds himself in debt. Anyone who incurs a debt in order to get out to hunt is unable to erase that debt.

When the welfare or unemployment check comes and is cashed at the store, a certain percentage is automatically applied against the debt. The only cash flow available is that from fox trapping. Fox furs are still valuable and, where available, are heavily relied upon. Now there is talk that the white man in Europe is planning to put a ban on the importation of all furs. Some 100,000 Inuit and Indian trappers in northern Canada are about to hear that it is cruel to trap animals and that animals are in danger from such trapping. They already know what the next step will be, and the tactics being discussed include some very sophisticated maneuvers.

Although the white man may look down on him, it is being recognized that the Inuit is not a simple primitive. He has learned that his enemy is not a harsh environment of cold and wind, but rather mere geographical and cultural distances from the sources of power. And these can be overcome. The tools are airplanes and media manipulation, negotiating skills, and an ability to solve problems in a simple, direct way.

North American Inuit may no longer care about defending their lifestyle in Germany, France, or the United States, or worry about trying to sell anything to Europe or North America. The markets of Japan, Korea and China are far more promising and those buyers are not impressed by Greenpeace, the International Fund For Animal Welfare, or any other group which values animals more than people.

The hunters of Greenland have taken a different approach; they have talked their government into appropriating a great deal of money which is being used to advertise their plight throughout Europe. Large tour buses carry native people, their dogs and sleds and costumes on a campaign to meet their consumers and their admirers. They arrive at a community, stop and put on a show. Costumes, dogs, music, all have a tremendous impact.

When the dancing and the show is over, the Greenland Inuit ask the thickly-assembled European audience: "Why do you want to kill us?" The shocked reply that no one wants to kill them is answered with an explanation of the effect of the pelt ban. It is pointed out that the people and their culture cannot survive without an independent source of income. That source dried up when the market for sealskin was cut short by legislation and public opinion which keeps dealers from handling any seal products. Even the tusk of the narwhal can no longer be sold in the Common Market.

The Greenland government's five-year plan is to introduce its people's plight to the European voting public. It is an important step in correcting what is, for northern people, an impossible situation. The tactic may or may not work, and is an experiment to be watched closely. The Canadian Inuit also enjoy immense popularity and a sort of "heroes of the north" mystique in Europe, yet the public has been largely unaware that they have suffered the same problems as their Greenland cousins. Canada has not attempted a public relations tour for its own sealers. Policy may now be to wait and see if the Greenland experiment has any effect.

The organizations within the protest movement have always been careful to make light of any detrimental effect which their actions might have on the lives of native people and ordinary fishermen. Some, such as Greenpeace, have claimed that there was never any intent to put an end to native subsistence sealing, although even Greenpeace has stated that native sealing for profit should be ended.

Of all those people who hunt the seal, the Newfoundlander has been maligned the most viciously and has been portrayed internationally as the cruel brute who clubs baby animals. It is no comfort to him to chalk it off to bad luck that the world finds harp pups appealing, or that he exists in a time when television can show the world how he makes a good share of his living.

Animal rights protestors in London are often seen on television, portraying the Newfoundlander: a man is frenziedly and loudly kicking and clubbing a simulated seal pup on the sidewalks outside Canada House, splashing "seal" blood and demonstrating to all who will watch that this is an "inhuman" act. The blood circus is a crude tactic, but unforgettable.

People in the US are not treated to this, but a news programme in Montreal or Ottawa or St. John's is very apt to include each London blood dance. It is news which Canada can't ignore, and to say that it hurts and diminishes everyone is an understatement.

Perhaps this is one reason why Canada's Inuit and Newfoundlanders are looking to the far east for seal product markets. They may feel that any attempt at culture change in Europe in this atmosphere is a hopeless waste of time. Their government has given them no public relations money to match that of Greenland, and it appears that for now all Europe is entirely convinced that it was right to put an end to the pup hunt.

Newfoundlanders realize that it was the whitecoat hunt which put them all out of business, yet to condemn it themselves in principle goes against the grain. The whitecoat hunt is defensible, they say, even though few outsiders believe it. They point to the policy statement of the World Wildlife Fund which defends the hunt as long as it is conducted in a humane fashion, for it does not endanger the herds: the World Wildlife Fund is of the opinion that an annual take of eighty percent young-of-the-year seals, and twenty percent "older" animals is a sound conservation measure, compared to one which would take a higher percentage of breeding stock.

Several hundred Newfoundlanders who worked on the commercial ships to take whitecoats have been used by protest to crucify all sealing. This has been easy because they have an ugly-appearing profession. The truth regarding sealing is not attractive to the public, and is not accepted at this time. It hasn't mattered that the whitecoat hunt has been proven humane, or that it is known to be a sound conservation measure. The public "doesn't like it when they kill those seals."

In 1984 the IFAW told the public that the money fishermen earn during their net fishing season is used for buying more clubs, knives and guns to support their sealing industry. Therefore, the intent of the fish boycott was to prevent this economic chain from working. Americans who loved the seals would refuse to buy Canadian fish.

The government of Canada has never been sure how to handle the problem of public misconceptions of sealing. In recent years it has turned from reliance on diplomatic negotiations abroad at the ministerial and legislative level, to use of an impartial body of experts which would be in a position to assure the public that sealing was being well monitored and thoroughly investigated, and has passed muster on all counts.

A first attempt was the formation of the Committee on Seals and Sealing, composed of non-governmental people, as an advisory body to the Minister of Fisheries and Oceans. This body was composed of marine scientists, humane experts, and representatives of the sealing industry. Members were not paid for their work, as they were not employees. The committee was as independent as possible of governmental red tape and politics.

The Committee on Seals and Sealing did a respectable job of observing sealing methods, and investigating the sealing industry and the protest against it. Yet it was not effective in convincing the public that the clubbing of pups was humane, and that quotas were not so large as to curtail herd growth.

The counter message, coined by Davies, Greenpeace and others, was continuously given, while reports of the Committee were only sporadically announced. The secret of the protesters' credibility and Canada's lack of credibility lay in Canada's failure to push the professional conservation message with vigor to match that of the opposition.

Perhaps this was because Canada's decision-makers erroneously assumed that the truth needed to be stated only once. To do so more often would be unseemly and might demean the attempt by bring it down to the "unprofessional" level of those protesters "who were in it for profit." (Although the business of protest made millions with its credibility marketing programme, this was never recognized as a tactic which must be emulated to avert disaster.)

This use of an objective, professional, conscientious body of competent people did not have the desired effect of bringing respect to the Canadian practice of commercial sealing.

In 1982 Canada and the European Economic Council jointly convened a working group of the International Council for the Exploration of the Sea. This body examined all data concerning harp and hooded seal population status in the western Atlantic and concluded that the harp population appeared "likely" to be growing. Hooded seal population status could not be estimated with enough confidence to satisfy the group that conclusions could be made about its growth or pup production.

Canada claimed that the report of the ICES panel was proof that its management plan for harp seals had been correct and effective, but the protest groups which had made a point of the "numbers game" decried either the report or the official interpretation of it.

The ICES report was interpreted for the public through a Canadian press release which was subsequently attacked with vigor by Barry Kent MacKay, commentator in Toronto's *Sunday Star*. Silence from government followed and the hoped-for effect on public opinion was diminished again.

Most members of the public would never read the report for themselves, and criticism of it would remain in the collective memory. Most would never know that MacKay had been a member of the Unionville chapter of the Animal Protection Institute. API educators in California had already demonstrated their concerns about the danger to harp seals with the "Threatened Wildlife" poster, which pictures them alongside truly endangered species.

MacKay never claimed to be an objective commentator. His views on sealing were always vitriolic. His public may have been unaware that he belonged to a vested interest group which had definitely presented harp seal population status with extremely low, misleading and inaccurate population estimates.

Again, one must be reminded that any information being presented may be accepted by the public if it is given in an aggressive way without the distraction of an immediate, consistent refutation of every point which is at variance with it.

In this matter of the battle over which side would effectively control the information given to the public about the Seal Wars, the protest side has always won through aggression, repetition, and consistency of attack. Truth, in and of itself, has not been an effective weapon, for it has seldom left the Canadian muzzle with enough velocity and follow-up to hit home.

In 1982 Canada attempted to convene a second panel of internationally-respected experts who would produce a public report on their findings. Canada proposed to the European Economic Council that experts in the field of humane treatment and killing of animals investigate the humaneness of sealing methods currently in use. This move was made because Canada anticipated that the EEC would soon recommend that its member nations ban the import of seal pup pelts.

The international panel on humane killing was never convened. The EEC knew its public would not accept the results of such a study. Davies' massive newspaper advertising campaign had caused enough people to contact their EEC representatives to make them all very uncomfortable about having to deal with a study which might find sealing humane.

In its refusal to convene such a panel the EEC was, in effect, admitting to the world that truth was not as important as public opinion. It did not wish to be forced to present its constituency with unpopular facts. This attitude was the result of a belief that the public would never change its mind about sealing. In 1982 the EEC saw no point in analyzing reasons why this belief had become so firm, or why it was so unshakable.

Canada made another proposal which the EEC also rejected: that the Common Market agree to the formation of an international commission which would be responsible for the total management of all seals in the northwest Atlantic. Rejection of this plan was a defensive move on the part of the EEC Parliament, which felt that the mood of its public was so adamantly against any sealing that it had better not approve any plans which might seem to sanction it.

Wildlife managers and professionals should realize that these attempts on Canada's part to justify her management of seals through research and selective hunting failed, not because the management plan was inept or incorrect, but because the public had been taught that *any* use of the resource was against its best interests, in collision with its value system, and that its own values were the only ones worthwhile.

The amount and manner of taking the resource had been the original foci of protest, but the final battle became a so-called *moral* issue. The *fact* of taking became the final definitive argument. It was based, not on any scientific evaluation of management or on data about humane killing, but on the fact that the public does not like seals to be killed by anyone for any reason.

Long established bureaucracies are perhaps the most conservative bodies known to man. Their patterns of action and reaction seldom change, even when elected officials come and go. Once again, in 1984, Canada convened an international panel of impeccable authorities whose duty it was to investigate all aspects of sealing and to recommend to Ottawa any changes it felt should be made, for any reason, in the practice.

The Royal Commission on Seals and the Sealing Industry in Canada was composed of seven members. Three were Canadians and the other four were from France, Great Britain, the United States and Australia. All were men who were leaders in their fields of professional expertise.

The Chairman was Justice Albert Malouf, a respected judge on Quebec's Court of Appeal. Dr. Ian McAllister, economics professor at Dalhousie University, and Dr. Wilfred Templeman of Newfoundland, former Director of Fisheries Research with the government of Canada, were the other two Canadians on the Commission. Dr. K. Radway Allen, a fisheries biologist from Australia; Dr. John Gulland, a British expert on marine resources; Dr. Patrick Geistdoerfer, research biologist from the Natural History Museum in Paris; and Dr. Russel L. Barsch, lawyer and former associate professor of business at the Washington Graduate School of Business Administration, completed the panel.

The press release issued by the Minister of Fisheries and Oceans which described the make-up of the Commission and its forthcoming role noted that "the presence of a majority of non-Canadians on the Commission should underline the study's objectivity and will provide an international perspective on the sealing issue."

Initial hearings would be in Montreal in late January 1985 and subsequent hearings were scheduled for Toronto, Vancouver, perhaps Newfoundland, Washington, D.C., and London. The Commission would "examine all the aspects of seals and sealing including the social, cultural, ethnical, legal, scientific and economic implications, research management, and internatonal comparisons." It would do the above by "searching for existing material, carrying out research work, inviting the public to submit briefs, and finally by holding public hearings to examine briefs or hear oral testimony." Canada declared that "there is no question that all points of view will be considered" in the matter of sealing within her borders.

News about the Royal Commission became common in Canada, but the American public was generally unaware that it had been convened. Hearings in Montreal got off to a disappointing start, the first day's proceedings marked by cancellation of speakers who claimed they were not yet prepared to testify.

The following days, however, drew a small collection of press and a full schedule of participants. A number of Inuit hunters and representatives of native associations, this writer, a representative of the fishing industry, a nutritionist who works in the Arctic, a member of Greenpeace International, a cultural geographer and an elderly economist all testified, as did Mark Small, President of the Canadian Sealers Association.

Hearings came to a close at the end of the first week with the announcement that the next session would be in Toronto. There, Dr. Dave Lavigne, a biologist who has always disputed the government's seal management policies, George Whitman of the Hudson's Bay Company, Tom Hughes of the Ontario Humane Society, and others, would all be heard.

It was expected that Greenpeace Canada would speak at the Vancouver hearings, and that Greenpeace USA would make its big push in Washington, as would other United States protest groups. The government of Canada would present its brief at a second Montreal hearing. There would be, eventually, an opportunity to present every view of the use of seals which had ever been expressed.

The Montreal hearings were held in the Palais de Justice. The atmosphere was austere and the rules were clear: no cameras, no tape recorders, no video equipment of any kind, and no audience participation or reaction would be permitted. Each witness was sworn in with hand on the Bible, and then presented testimony. At the close of each presentation questions were asked by the Commissioners. It was a no-nonsense courtroom procedure. All press representatives were instructed to do interviews and take pictures outside the room during breaks and at the end of the day.

Some witnesses were understandably nervous and afraid that they could not present their briefs in an effective, credible manner under these intimidating conditions. A great deal was at stake. Each Inuit representative, and Mark Small, reminded the Commissioners that he was unaccustomed to such an atmosphere and hoped that his message would nevertheless be accepted as representative of the feelings and intentions of his people back home. Each was assured that he should speak in as natural a manner as possible and that he would be treated fairly.

Those men who had travelled to Montreal to speak for their people back home came dressed in business suits, their faces showing strain. Each would feel responsible if his presentation were not received as significant or credible in the overall perspective of sealing which the Royal Commission was attempting to achieve.

Paul Okituk, researcher for the Makivik Corporation, was the first Inuit to speak on Wednesday, January 23. Okituk is a young, not very tall man with dark straight hair and those facial features which Inuit share. His presentation was meant to illustrate the importance which sealing still has for modern Inuit in northern Canada, even though "modern" ways and food sources have been introduced. He pointed out that he could have come "in full regalia" with a sealskin suit which would make everyone who saw it jealous, since it is such a magnificent costume. He said, somewhat shyly, that his mother had made it for him because he is not yet married: wives make clothes for their husbands, and he did not have one yet. He described the outfit and said that today the best and warmest clothing is still made from sealskin, although now many Inuit wear imported clothing because they cannot get out to hunt as often because it is now so expensive to go hunting and, finally, because the new clothing is lighter weight and easier to move around in. It does not, however, make a person as proud as does a sealskin outfit.

Okituk felt that the Commission should be aware that Inuit communities have suffered greatly since the ban on seal pelts which has affected the market for all fur goods produced in the circumpolar regions. He stated that, contrary to popular, "southern" belief, there is no pressure on those renewable resources such as seal and caribou which are routinely hunted.

He pointed out that circumpolar uses of the seal are, in order of importance, as follows: (1) meat; (2) bones are still used for tools and needles; (3) sinews are used for threads; (4) skins are still used for clothing and for cash. Sixty percent of the diet of the Inuit before the ban was composed of protein from marine resources but is generally less now, although, for people in his community, seal is still a mainstay.

Okituk felt it hard to convey a realistic picture of northern life to his southern listeners who are not used to harsh environmental and weather conditions and who expect to live long healthy lives. He said that life itself is a miracle in the Arctic and something for which to be grateful. The preservation of life is the strongest instinct we possess, he said, and it was obvious that he felt the Commission should realize how vigorously Inuit would work to preserve the quality of that life.

Although his people realize that there is a crisis of sorts which currently affects their lifestyle and their very existence as a cultural entity, he stated that there is a general optimism that this very bad situation must, in time, get better. After all, he said, we live in a very forgiving world. Did not the world forgive Germany for its Nazi era and did not the world forgive Japan for its imperialism? Okituk feels that the world will also forgive the hunting of seals, in time, and that the markets shall either return or more shall be found. He spoke of creating new markets in North America and in China.

Some of his listeners did not share his optimism, but all were impressed with his sincerity. This was a good man, and he spoke for a good people who have suffered without good reason. They feel no bitterness, but want life to return to normal as soon as possible.

The Commission questioned Paul Okituk as to the amount of seal use now and in the time before the ban, and thanked him for doing a fine job of presenting the native perspective. His judges were thus on record as having found his presentation one of significance.

For hours before it was his turn to speak, Mark Small had been very concerned about his own presentation. As President of the Canadian Sealers Association, he had to speak not only for the landsmen, but also for some who had been commercial sealers and for some in the Magdalen Islands who had traditionally taken whitecoats. He knew that everyone's trouble had stemmed from the publicity surrounding the whitecoat hunt and that the world knew only about that and nothing about the many thousands more and their families to whom the hunt for beaters and other older seals was an integral part of the yearly cycle.

Small knew that protest literature about the hunt had always minimized the number of people who depended on it, and the place which sealing played in the total economy. He was a man with limited formal education whose task it now was to represent in court thousands of people who would not have the opportunity to defend their way of life.

Mark Small was sworn in and began to speak, quietly at first, and even almost timidly. But after his brief introduction the anger and frustration began to take over. This was not a stranger in a gray suit talking to other gray suits about worldly economics and corporate strategies. This was a fisherman, fighting mad, from a world of boats and nets and unpaid bills, daily danger and economic risk. Those who sat listening to him, the Inuit friends of Paul Okituk still wearing their jackets, the press in their work costumes, the other sealers and the men of the Commission, all saw and heard a common man who spoke with an uncommon, passionate eloquence.

> I stand before you today as a sealer, as a fisherman, as an innocent man whose livelihood has been unjustly taken away, and whose reputation has been unjustly vilified.
> As a sealer, I look forward for our industry to be fully and objectively examined so that my son and my son's sons can, for generations, continue to harvest with dignity and pride this renewable marine resource.
> I never imagined that the day would come when I would have to come into a courtroom to state my right and the right of the thousands of sealer-fishermen to retain our livelihood. I've never been in a courtroom before to defend myself. There were eleven children in my family, we could not get a proper education, so I find it difficult to appear before you today. But I know that we must do this, because if not, then all of Canada's primary

> producers will be adversely affected by a movement which is attempting to redefine our relationship to the animal world.
>
> Now it is sealers. Next it will be our hunters and trappers, then all those involved in the fur industry and eventually our farmers, as this ill conceived and dangerous animal rights philosophy leads us towards a vegetarian society. We must draw the line now. As a sealer I want to have my voice heard in this debate which is vital to the future of our society....

He went on to outline the history of the formation of the Sealers Association and the way the issue of seal protest began to be recognized in Newfoundland:

> It was apparent to us fishermen that although the debate over the sealing industry had gone on for years, the efforts of politicians and government officials achieved very little headway. Worse, it appeared that every spring the debate in the media only served to further mislead the public and reduce the value of our industry. It was clear to us as sealers, the people who stood to lose or gain the most, that we had been left out of this debate.

He outlined the role which the Extension Service at Memorial University of Newfoundland had played in introducing sealers to the full impact of the movement against them, and the way the outside world viewed their tradition, then the role played by their university in showing them how important it was to create a new public image for themselves. In 1982 the Canadian Sealers Association grew out of this stimulus, formed as a defensive organization, even when most felt it was already too late. It began at Baie Verte, on the northeast coast, with a small group of desperate men collecting a few hundred dollars in a hat, and grew to become a representative, even a political, force, eventually subsidized by the federal government.

> The aim of our association was set out by fishermen to 'preserve, promote and protect the sealing industry'. This was to be done by educating the Canadian and world public; by working to change the image of the industry and by working in cooperation with both levels of government to develop new markets overseas and to promote a Canadian based industry.

They decided that a complete revitalization of sealing was the only worthwhile goal, and even though this seemed impossible to some, it was worth all the energy which they could put into it.

Their industry had been dead, almost from the moment the Baie Verte meeting was convened. Davies had prepared the European public with a three million dollar advertising campaign, and the result was disaster. By 1983 the pelt market had gone from acceptance of 178,000 (mostly whitecoat) skins to grudging purchase of 58,000 beater pelts. Mark Small, and many like him, saw income from sealskin go from $10,000 in 1981 to $184 in 1984, to absolutely nothing in 1985. His response was "Where do I look to replace that income?"

Brian Davies had been partly right. For Newfoundlanders and others, sealing is largely a matter of dollars and cents. He expressed this in 1969 as a simple problem which could easily be overcome with government subsidies of the sealing industry in the form of alternate industrial development (perhaps tourism) and increased unemployment insurance to compensate for lack of a market. Franz Weber offered to set up a stuffed toy factory so that Newfoundlanders could turn from killing to making effigies of their prey, cute toys for children, stuffed by stiff fishermen's hands which had never worked with fuzz and ribbons and sawdust, sewing machines and factories. Weber and his world never understood why the idea didn't catch on.

The sealers of Canada did not want to become factory workers, making toys for export. And they did not want to join in the spirit of the anti-sealing debate and call their opponents names. Their desire was to work quietly to develop their industry, not to defend it out of anger. They felt that it would be a waste of their time to attack the leaders of anti-sealing groups with defamation campaigns.

The desires of those Canadians who fish and hunt seals for a living are simple and their goals are to continue life in the old way, through use of seals and fish: "We are proud primary producers. We are the people most dependent on animal resources for our livelihoods. As such, we have the greatest respect for these living resources. We are practicing professional ecologists."

The fisherman stood before his judges and listed his fears and hopes, his traditions and habits.

> Mr. Chairman, it makes me very sad and very angry to hear people say that we should be on welfare. Our people do not want government handouts. We do not want to be a burden on Canadian taxpayers. We want to continue earning our living with pride and dignity. In good years with the right ice conditions we can continue sealing right through until the end of April month. The money that we earn from this part of the fishery is absolutely critical for us to continue the rest of the year's fishery. About the same time that the seal fishery ends we bring our boats on dry dock for their annual refit. If there is a poor seal fishery we will not have the funds necessary to repair our boats properly, to invest in new equipment, to buy new nets, and to make

payments on our loans. Without the seal fishery, we cannot get the rest of the year's fishery off to a good start.

Our fishing season is often shortened by a month at either end through harsh weather and ice conditions. In the fall of the year most freezers in homes throughout rural Newfoundland have a wide variety of wild foods, including berries, rabbits, various wild birds, moose, caribou and, of course, seal meat.

If we are told that it is morally wrong to kill a seal, then it must be morally wrong to kill a moose to provide food for my family. To survive we must use all of the natural foods which are available to us.

It is not a well known fact, but it is accurate that the great majority of seal meat is fully utilized. It angers me when I see on the TV pictures of whitecoat seal carcasses just left on the ice. There is very little useable meat on an animal of that age. The flippers are used, but the TV coverage doesn't show that. On the older animals which we take, all of the meat is used.

We know that many people living in urban areas would covet our lifestyle if they only knew that it existed. We are poor in many ways. Yet in others we are rich.

We survive month to month, year to year, living in hope for better times. On average, our incomes are well below the poverty line, yet we live a lifestyle that brings great day-to-day satisfaction. We have often heard from our critics that men such as myself only earn a few hundred dollars a year from sealing. Therefore, it is of no great economic benefit. But Canadians and this Royal Commission must realize that for families living near the poverty line, a few hundred dollars means a lot. Without that money we can't continue to make money, because we need it to reinvest in the rest of the year's fishery.

There are a large number of fishermen, however, who derive many thousands of dollars each from the seal fishery. Men such as myself who put a substantial amount of time and effort into it can earn up to a third of their yearly income from sealing. We are insulted by the arrogance and ignorance of those people who criticize our industry on the basis that it has little or no economic value to individual fishermen.

Without the seal fishery our forefathers could not have settled our coast. Without the seal fishery our lifestyle will be gravely threatened. Without the seal fishery my livelihood will be seriously jeopardized. I am a small businessman. My industry is in deep crisis. Little by little there is an inevitable domino effect. We see the signs around us every day. The fishermen are losing their cash flow and their line of credit. We are all technically bankrupt, unable to earn enough income to replace our capital investment in equipment.

Sealing by itself is a small industry, but it is an absolutely integral part of our commercial fishery and subsistence living. In all of this there are occupations and responsibilities which create a healthy, vibrant and stable community. Take one element out, begin to erode the lifestyle of our people, and you will see economic and social collapse. Already I see it in my own community.

The ice is now off our coast. Hundreds of thousands of seals are out there. The boats are tied to the wharf. We have to stand on the shore and look out to the sea.

This time of year there should be lots of activity. Instead of the joy of returning to work again in the fishery there is a quiet despair and desperation which hangs over us. We are saddened. We are angry. We are fearful that this very important part of our life could be lost to us forever if we don't stand up and make our voices heard.

This year when I begin my fishing season in May, I will see the seal not as my friend who helped to provide me with a means of subsistence through the winter months, but as a competitor for the fish which I need to survive. Although we know it will be a long time before it is documented by government scientists, we live close to these animals and we see signs of an increasing population already. Seals are now getting tangled in our nets in places where they never were before. Seals die needlessly. In the process we lose the fish which were in the nets.

We are told by the government that the mature seal eats more than a ton and a half of fish per year. The total consumption of seals is more than the total allowable catch of Canadian and foreign fishing fleets. We know that our present scientific knowledge of the intricate ecological relationship between all forms of marine life and their interaction with seals is imprecise. But we plead with our scientists and with this Commission to include fishermen as a necessary part of the ecological equation. Surely, government will not wait until there is a substantial loss in our income from a reduction in fish stocks before a cull of seals is condoned. I am not a scientist. I am not a biologist. But to me as a fisherman, it is just common sense. I see the seals eat turbot, caplin, herring, and crab. Fish that I need to survive. Surely we cannot sit idly by and let nature work out its own balance of all these species while thousands of fishermen collect welfare.

What is our future? What is the future of our sealing industry? This year for the first time I will not take my longliner out sealing, except to get meat for family and neighbours. I and thousands of fishermen like me have too much pride to let our industry go. We want to survive by the work of our own hands, with the pride and dignity of an earned income. We don't want handouts. We want more jobs. We don't want those we already have taken away from us. We don't want to be a burden on the taxpayers of Canada. We want to contribute to the economy. We want to make sure that this crisis in our sealing industry becomes an opportunity to create more jobs in the communities where unemployment can be as high as eighty and ninety percent.

In the sealing industry, as with many other Canadian industries, we have been hewers of wood and drawers of water. We have not, in the past, taken steps to develop our own industry across Canada. We feel that new jobs can be created, not only for fishermen, but for other people who can make sealskin products for distribution throughout our province, Atlantic Canada and abroad. If we are successful, this will be an important development not only for its economic value but because the wearing of Canadian-made sealskin products is a statement of support for the industry. It is important that all of us involved in this industry have the opportunity to display the pride that we take in our livelihood.

Man's role on this planet is not to preserve what nature provides, but to conserve it. Seals and all the other species we depend on are renewable marine resources which were placed there by God with a clear message to man. We not only have a right, but a responsibility to utilize our living resources wisely so that they will be there for generations to come.

As sealers, we believe that the seal fishery must be practiced on sound scientific principles to ensure that the overall seal population is stable, healthy and increasing in size at a controlled rate, to ensure that the harvesting methods are as humane as possible, and to ensure that full and practical utilization is made of the entire resource, including the pelt, the meat, the fat and other by-products.

The seal hunt is practiced with respect to these basic principles. Its continuance is therefore justified on both economic and ecological grounds. We can be proud of the seal hunt as a model of excellence in wildlife resource management. There are many examples of our society's insensitivity to the ecology of which we are a part. The seal hunt is not, however, one of them.

I am not a barbarian. I am not a savage hunter. The accusations of cruelty and inhumanity have had a deep effect on fishermen and their families. We do not understand how anyone could say such things.

Our sealers association is in the process of completing a professional, thorough, and objective national survey which assesses Canadian attitudes towards the sealing industry and sealskin products. Early returns from this study are most encouraging. When asked if killing of wild animals is acceptable if a person's survival and livelihood depends on it, over ninety percent of Canadians agreed. Over eighty percent of Canadians agree with the statement that seals should be taken or killed by the most practical method, provided it is done humanely. It is also apparent from this study that a great deal more of education work needs to be done. Only fifteen percent of Canadians understand that harp seals are not an endangered species.

Canadians do not know who the endangered species really is. As a sealer, as a fisherman standing before you today, I say to you that I am the endangered species.

I am endangered, but I will fight back and I will survive. I will not let animal rights become more important than human rights. I will not let people give souls to animals while they rob me of my human dignity and my right to earn a livelihood.

Yet as a true Christian I forgive these people who have made such statements. We must enter into a new relationship, respecting that we all care about the environment. We must all work together to ensure the proper maintenance of that environment. The future of us all depends on it.

I will stand tall and proud along with my son and my son's sons who will be sealers for generations to come. Thank you.

The fisherman was questioned briefly, and returned to his seat. He had done his best, and the room was silent. A number of people, both men and women, found themselves moved to tears, for they realized that he had described the crucifixion of a unique, cherished way of life, had even shouldered some of the blame for it, for not having realized in time that the general public should have been educated sooner. He had spoken of hopes for a resurrection which in the opinion of many was an impossibility.

The next person to be sworn in for testimony was John Amagoalik of Frobisher Bay, Northwest Territories. This Inuit man is tall compared to most, and very slender. He wore a gray suit, as did his friend Peter Ernek of Rankin Inlet. He and Peter would jointly present information to the Commission on the importance of seals and sealing to their communities, and their concerns about the impact of a general ban on all sealing.

Amagoalik is a well-spoken, quiet, serious, credible representaive of a desperate people. His topics included the use and importance of seals, aboriginally and in recent history. Current levels of use and the importance of seals as a source of cash income in a changed economy were stressed as strongly as they had been by Paul Okituk. He mentioned that cash was necessary in order to hunt, and that seal meat was so vital to the Inuit diet that lack of it was becoming a significant public health issue.

And, for an Inuit, to hunt the seal is as important for social reasons as for the material comforts it brings. This is the way in which men pass on, not only important technical skills, but social knowledge and appropriate behaviour to their children.

> For example, it is through the hunting of seals, and their butchering and distribution, that young people can readily be taught the virtues of cooperation, patience, sharing and their responsibilities in the community.

> Similarly the processing of seal skins, oil and meat is an important means by which women pass on similar cultural traditions.
>
> Sharing is an important part of the Inuit ethic, and in some parts of the Arctic, seals are the chief product which is shared among families and even sent from one community to another, to reinforce the bonds of solidarity among relatives.
>
> Sealing, or any kind of hunting, is not the same to the Inuit as it is to people down south. It is not simply recreation, or a blood sport. It is the central means by which the Inuit afirm life, and for many, indeed, still is a part of everyday life. Most certainly it is neither a pastime for the rich nor an occasion to get out once a year and drink with the boys.

Amagoalik also pointed out that a ban on importation of seal pelts into a European market would not put an end to the hunting of seals. An established way of life dies hard, even when it becomes prohibitively expensive. He knew that some protest groups were calling for a complete ban on all sealing everywhere, and wanted to make sure that the Commission understood what the consequences would be for his people.

He spoke of the breakdown of social order in northern communities, the higher incidence of suicides, violent deaths, alcohol and drug abuse, child neglect, and other indicators of social pathology which affect those areas where the natural resources are suddenly denied the people. This had happened with disastrous result when the Ojibwa at Grassy Narrows and Whitedog in northwestern Ontario had suffered the complete loss of their fishery due to mercury pollution in 1970. Both the chief source of food and the chief source of income had been taken from them due to the pollution from modern mining activities.

John Amagoalik felt that the loss of the right to hunt seals would have the same devastating effect on the Inuit. He noted also that a ban on Arctic sealing would inevitably hurt other animal populations, such as the caribou, whales, geese and anadromous fish which are "neither as ubiquitous nor as plentiful as seals." Having to rely on those species as substitutes for seal would cause overharvesting and depletion. He noted that the Inuit "rely on many food species in balance, and removing access from one of the main ones would throw everything else out of balance."

Although the technology of seal hunting has changed with the introduction of guns and gasoline powered vehicles, the impact on the population has not altered. Seals up there are still thriving. They are still hunted as individuals. Since the methods have remained the same, "the impact of Inuit hunting on seal populations has hardly changed since aboriginal times."

> The Inuit have never been the cause of seal population declines or local extirpation in the Arctic, either by virtue of their hunting techniques or the fact that sealskins are sold commercially. What new technology has really meant to the Inuit is that they can maintain an ancient system of seal hunting while living together in modern communities instead of on the sea ice in igloos.
>
> The Inuit hunt is humane. Seals are normally killed by direct head shot from a high-powered rifle. This is probably the most instantaneous and least painful of any of the ways that a seal can die from human or natural causes.
>
> The Inuit share the concern of others about the real dangers to the health and abundance of seal populations, which are especially from improperly controlled oil and gas exploration, development and transport, and the possible widespread pollution of the marine environment. The Inuit have repeatedly addressed their concerns in public hearings and other forms, in alliance with conservation organizations. But the Inuit have not been content merely to talk and to advocate. In the course of negotiating their claims settlements, they have devised a practical, workable system of fisheries and wildlife management and conservation in cooperation with the governments of Canada and the Northwest Territories. We invite those governments to fully implement this system, and we invite all others interested in a healthy and abundant seal population to work with us to that end.

The hunter than gave his opinion of the danger of those views which oppose the seal hunt.

> The Inuit do not disagree with those who promote the conservation of seals and of their marine environment, the humane hunting of seals, and who are against the wasteful use of seals or any other wildlife. However, the Inuit are very concerned about what seems to be the underlying view of many of those opposed to the seal hunt, which is that hunting of any kind is wrong, and indeed the human use of animals in any way is wrong. The Inuit have always depended on animals for life, because they do not and cannot rely on agriculture.
>
> The Inuit culture requires hunters to behave respectfully and responsibly towards the animals they hunt, and not to misuse their power with respect to the animals. The Inuit recognize that this relationship to animals does not exist for many people in the industrialized regions of the world, and if people in those areas are critical of man's attitude to animals there, the Inuit believe they have a right to be critical. But people in those areas

certainly do not have a right to tell the Inuit how to live with animals, because the Inuit need lessons from no one in that regard. Other people may feel themselves separate from or alienated from the natural order, but we do not.

The Inuit believe that a proper and respectful use of animals is essential to life. If people in the industrial cities are allowed to stop others from far away from hunting, then what is next? Will they ban fishing? Will they ban the keeping of cattle, pigs and poultry for meat, eggs and leather? How will people eat? The Inuit see the struggle to maintain the seal hunt not in isolation, and not just a matter for their own welfare alone. This is why the Inuit have allied themselves with other seal hunters, and in a larger way, with all who obtain and use animals properly and respectfully, because this is clearly how mankind must live. The Inuit are deeply concerned about where this extremist philosophy about the seal hunt will lead to in the future.

Mr. Amagoalik then turned from the panel of Commissioners and from his microphone to face the sealers of Newfoundland who were sitting in the front row behind him. He gestured farther back at the three empty seats where the representatives of the International Fund For Animal Welfare had been consistently parked in the days before. Today they were conspicuous by their absence, perhaps not wanting to hear the testimony of those for whom they had caused so much fear and grief.

John Amagoalik was through with his people's printed presentation, but he had one more thought to offer. He spoke quietly, as one hunter to all others: "and we Inuit, along with the Christian fishermen of Newfoundland, want to say that we also forgive them, for they do not know what they have done."

Postscript

In the weeks to come the Royal Commission on Seals and the Sealing Industry in Canada would hear and read the testimony of hundreds of people regarding their stand on the use of seals as a resource. The entire body of recorded oral testimony and thousands of pages of written briefs would be examined and weighed for the purpose of deciding what the Commission's recommendations to the government of Canada would be.

In the meantime, in the Gulf of St. Lawrence and off the northeastern Newfoundland coast, the seals would come onto the pack ice. Seal time would see the births of some 500,000 white pups of which perhaps a few thousand would be taken by people for meat, fat and fur.

In February 1985 Phil Donahue's television programme (*Donahue*) in the United States featured a panel of people, including Peter Dykstra from Greenpeace USA, who abhorred the use of seals, whales, cattle, poultry, swine, fish, laboratory animals, and any other creatures, wild or domestic, for consumptive human purposes. They agreed that it was morally wrong to take animal life for meat, leather, fur, or medical experimentation of any kind because this is cruel, unnecessary, not demonstrably beneficial, and wasteful. Although the studio audience did not appear convinced that the panel was credible, the fact that the host and his producers gave them spectacular media time was a victory for them all.

The Greenpeace position has apparently changed radically from that which Patrick Moore had outlined up through 1983. During that time he discussed a philosophical and moral contrast between the use of domestic stock for food and leather, and the comparable use of any wildlife. Since then, however, Dr. Moore (a director of Greenpeace International) claims to have become a vegetarian.

The alliance in 1985 of Greenpeace USA with a totally vegetarian position has been a surprise to those who keep an eye on the direction of the protest industry. Although a switch from selective to total abstinence of the consumptive use of any animals is a logically consistent one for Greenpeace to have made, no one really expected it to be a part of organizational policy. At least, not so soon.

Why would Greenpeace opt for such a radical position? One might think that this would turn off many of the "general public" who support the movement. It could be speculated that perhaps having achieved their ostensible goals of ending large-scale whaling and sealing Greenpeace might have been afraid they had put themselves out of business. Of course, it is still an imperfect world!

There are still cruise missiles, nuclear submarines, nuclear power plants, hazardous waste and acid rain to demonstrate against, so the organization shall probably keep the loyalty of those who share those concerns. Greenpeace will continue to attend the CITES and the International Whaling Commission meetings.

There is the possibility that Greenpeace's public relations advisors felt that the group could sustain its base of international support with the new, logically consistent stand against the consumptive use of all animal life, while at the same time pursuing the old environmental issues. After all, Greenpeace would be simply setting a "moral" example, not necessarily calling for the world to join them in this vegetarianism. In this way, the organization might pick up members from the radical animal rights segment while retaining most of the old list of supporters. Those of the general public who ceased to support Greenpeace might be more than compensated for by the recruitment of the new "radical fringe" animal rights constituency. What appears on the surface as a move of philosophical solidarity with those elements, may be a pretty shrewd way to seduce a previously untapped but growing market.

As for the sealing issue, Greenpeace policy has always stated that there is no intent or desire to harm subsistence hunters. The problems in the '80s lay in the definition of "subsistence" sealing. Instead of trying to find a financial ceiling of dependence at which these outsiders might draw the line, and instead of fighting the growing tendency of the media to portray all sealers as victims of human rights violations and of unfair outside cultural discrimination, the organization appears (as stated on *Donahue*) to have decided that even subsistence hunting is wrong: sealing is bad, whaling is bad, Big Macs are unconscionable and really good people should give up all meat and wear plastic shoes.

As for the IFAW, it is not known if it will continue to protest sealing internationally since most of the public think the issue is dead. Steve Best's focus of interest changed sometime during 1984, as he became a representative of the I Kare Wildlife Coalition. It is not known if he also remained a "consultant" for the IFAW. Best claims that I Kare has been in existence since 1973; in 1985 he was its vice-president.

In January 1985 Best published and circulated a soft cover document entitled "The Royal Commission on Seals and the Sealing Industry in Canada: The Perceived Biases of the Commission." This paper states that the Commission was convened by Canada only as an attempt to influence the next vote of the European Economic Council on the question of extending or lifting the ban on the import of "baby" seal products.

Best claims that the data search which he had conducted on each commissioner reveals that each is either demonstrably incompetent to discuss seals or is definitely, according to previously published statements by each man, biased on the side of wildlife management. Thus, according to Best, it is apparent that the commissioners were predisposed to rule in favour of Canada's past and present management plans for seals. He did not list alternative natural scientists or other professionals whom he felt would have been more appropriate as members of the Commission. He concluded that, because of this alleged bias, it was useless for protest organizations to bother to testify or submit briefs.

It does not appear that his call for other groups to boycott the Commission was a success. There is no denying that Best was astute in his assessment of the probable outcome, however. The competent marine scientists, the economist and the lawyer with experience in the native rights movement are certainly experts able to consider rationally the matters up for discussion. No unfounded argument, protest or otherwise, would be likely to snow them under. The vested interests of I Kare, IFAW, Greenpeace, HSUS, API, or any other animal rights or environmentalist group would definitely be publically jeopardized by the objective report of such a body, if their past actions and policies had been in disregard of the facts. The main danger these interests would face would be their loss of credibility from this source. Therefore, the credibility of the Commission had to be undermined. It was, in retrospect, a predictable defensive move.

Best's very thick document was carefully circulated among press, writers, the Commission itself, the Department of Fisheries and Oceans, and other protest groups. It may well be compared to his original fish boycott paper for its apparent intent to demoralize, frighten, annoy and harass his opposition.

Culture change on a large scale is normally a slow process, aided by fluid communication networks among differing groups. It is possible to demonstrate slow and steady change in a culture through the smoothly changing frequency curves of the acceptance of new elements of belief and usage. Now that the human rights issue has been recognized by the public as a valid part of the sealing debate, there is some evidence that such a process of change is taking place in thought patterns in the western world. Hints of this are seen in European attitudes towards the Greenlanders, and in North America where the plight of the Inuit and Newfoundland sealers has slowly been brought before the public.

It is possible that there is already in motion a new pendulum shift in public perception of Greenpeace, the IFAW, and other groups which have participated in this movement to curtail or end the use of animals, either wild or domestic, by man.

In the opinion of this researcher, there is no real justification for drastic curtailment or abolition of the harp seal hunt, and no reason why legal hunting of any other seal species in the north Atlantic should be curtailed or ended. This perspective is shared by all non-vested interest observers of the hunt who originally felt that the humane and species welfare issues should be investigated and addressed.

One has to come to the conclusion that, because of the misinformation purposely presented to the public in the Artek film, original investigations of hunt procedures were justified. It has been recognized that there *was* a need for continuing research into the impact of various levels of harvest on a large population of harp seals. This has all been accomplished. Research has continued throughout the entire period since 1949, when Newfoundland became a part of Canada.

The public can be assured that protest descriptions of hunt impact on seal herd status are incorrect, as are claims of cruelty, both physical and psychological, to seals. The reasons why organizations of protest continue to "harp" on this are unclear. It is the opinion of this author that misinformation is still being given, even in the face of all scientifically supported data to the contrary. One must assume that most of these data have reached the administrators of such organizations. They are either chosing to ignore them, or are continuing to declare them invalid. Alternate specific interpretations, however, have not been offered to the scientific community in rebuttal.

There is now an impasse in this matter. One might legitimately ask why, if protest organizations are truly concerned with cruelty and species impact, they have not offered compromises to procedures and harvest levels. This would seem to be a reasonable course. Instead of this, protest organizations have consistently called for a total end to the hunting of all seals, while denying the adverse impact that this has had on those who depend on the resource. Lack of adverse impact on seal herds has also been consistently denied by protest literature, which has continued to claim that seals are endangered by *any* harvest.

Consequently, there has been a growing belief among seal hunters and marine scientists alike that the forces of protest are primarily concerned with maintenance of their credibility with their large constituencies, and now find themselves locked into a position of being unable to retract their unjustified claims and goals. This explains why, for many, the seal hunt issue had come to be a "moral" one. Cruelty has been disproven. Adverse herd impact has been internationally refuted. "Moral" reasons why seals should not be taken are the only ones left.

It is now being recognized that protest-caused forced culture change is itself morally bankrupt as a course of organizational direction. The impact of these organizations on the well-being of Native and Newfoundland hunters, and on others who regularly use wild or domestic creatures for food, clothing, sport, transport or research, may be lessening somewhat through a general decline in their credibility. The many victims on both sides of the seal wars, and those involved in a number of other animal-use and so-called environmentalist campaigns, are waiting to find out.

Bibliography

Animal Protection Institute of America. "Harp Seal Update." Prepared by API staff Penny Feltz and Narca Moore-Craig. January 1982.

Beddington, J.R. and H.A. Williams. "The Status and Management of the Harp Seal in the Northwest Atlantic. A Review and Evaluation." PB206105. Washington: U.S. Marine Mammal Commission, July 1980.

Best, Stephen. "The European Economic Community, Canadian Fish Products Boycott." Distributed by the International Fund For Animal Welfare, Canada, July 1982.

_____ . "The Royal Commission on Seals and the Sealing Industry in Canada — The Perceived Biases of the Commission." January 1985.

Bowen, W. Don, C.K. Capstick, and D.E. Sergeant. "Temporal Changes in the Reproductive Potential of Female Harp Seals (*Pagophilus groenlandicus*)." *Canadian Journal of Fisheries and Aquatic Sciences*, Vol. 38, No. 5 (1981), 495-503.

Bowen, W. Don and D.E. Sergeant. "Mean Age at Sexual Maturity and Age-Specific Pregnancy Rates of Northwest Atlantic Harp Seals in 1980 and 1981." Northwest Atlantic Fisheries Organization NAFO Scr. Doc. No. 81/XI/152 Serial No. N459. Special Meeting of Scientific Council, November 1981.

_____ . "Further Estimates of Harp Seal Pup Production Between 1977 and 1980 from Mark-Recapture." Northwest Atlantic Fisheries Organization NAFO Scr. Doc. No. 81/XI/152 Serial No. N459. Special Meeting of Scientific Council, November 1981.

Canada. Department of Fisheries and Oceans. "The Atlantic Seal Hunt: A Canadian Perspective." 1984.

_____ . "Critique of the Nature Conservancy Council Report of 1982 (Annex II)." Presented to the 1983 CITES Convention meetings, Botswana, April 1983.

_____ . "Instructions to Fishery Officers on Annual Seal Patrol." March 1982. Internal memorandum.

_____ . "Official Report of Incidents which took place during the 1981 Seal Hunt off Prince Edward Island." 1981.

_____ . *Questions and Answers on the Seal Hunt*. January 1979. Brochure.

_____ . *The Sealer's Guide*. January 1981. Manual.

_____ . *Sealing Instructions For The Humane Killing of Seals*. 1980. Brochure.

Canada. House of Commons, First Session, Twenty Eighth Parliament, 1968-69. Official Bilingual Issue. Standing Committee on Fisheries and Forestry, Minutes of Proceedings and Evidence. Nos. 13, 14, 21, 24, 25. March through May, 1969.

Canada. Royal Commission on Seals and the Sealing Industry in Canada. "Appendix 'A' — Statement of Policy and Procedure." Adopted at the first meeting of the Commission. September 1984.

Canadian Sealers Association. "Presentation by the Canadian Sealers Association to the Royal Commission on Seals and the Sealing Industry in Canada." Presented by Mark Small, President. Montreal, January 1985.

Davies, Brian and Eliot Porter. *Seal Song*. New York: The Viking Press, 1978.

England, George Allan. *The Greatest Hunt in the World*. Edited by Ebbitt Cutler. Montreal: 1969.

Hold, Sidney and David Lavigne. "Seals Slaughtered — Science Abused." *New Scientist* (March 1982).

Hughes, Tom. "Report to the Committee on Seals and Sealing — 1983 Seal Hunt." 1983.

International Council For The Exploration of the Sea. "Report on the Meeting of the Ad Hoc Working Group on Assessment of Harp and Hooded Seals in the Northwest Atlantic — Jointly Convened by Canada and the EEC Commission for scientific advice from ICES on aspects of the population dynamics and state of harp and hooded seal stocks in the Northwest Atlantic." ICES Headquarters. October 1982.

International Fund For Animal Welfare. "Would You Take a Simple Step." Fund literature to public asking for compliance with the fish boycott. USA mailing. 1984.

Jewell, P.A. and S.J. Holt, eds. *Problems in Management of Locally Abundant Wild Mammals*. New York: Academic Press, 1981.

Lavigne, Dr. David M. "Effects of Dye on Harp Seal Pups." *Behavior and Energetics of Whelping Harp Seal in the Gulf of St. Lawrence*. Research report prepared for the Department of Fisheries and Oceans, Ottawa, 1980.

Lederer, William J. and Eugene Burdick. *The Ugly American*. New York: W.W. Norton & Co. Inc., 1958.

McCloskey, William. "Bitter Fight Still Rages Over The Seal Killing in Canada." *Smithsonian Magazine*, Vol. 10, No. 8 (November 1979).

_____ . "Killing Baby Seals: A Sympathetic View." *Washington Post*, September 15, 1983.

_____ . "Harp Seal Hunting." *Oceans* (November 1983).

Nationwide Market Research Corporation. "Attitudes Towards the Harpseal Hunt — A Research Report Prepared for the International Fund For Animal Welfare." September 1982.

Nature Conservancy Council. "Recommendations and Status Reports on Harp and Hooded Seals — Revision of the 1981 Report Eur 7317 EN." Prepared for the Environment and Consumer Protection Service of the Commission of the European Communities by the Nature Conservancy Council of Great Britain. Contract No. u/81/525. May 1982.

North Atlantic Fisheries Organization. "Stock Assessments of Harp Seals." NAFO Scs. Doc. 80/XI/34 Serial No. N263 p. 3. Special Meeting of the Scientific Council, Dartmouth, Nova Scotia, November 1980.

Northwest Fisheries Organization. "Assessment of Seal Stocks." In "Scientific Council Reports." Dartmouth, Nova Scotia, December 1983.

"Panel on Euthanasia Report." *The Journal of the American Veterinary Medical Association*, Vol. 173 (1 July 1978).

P.J. Usher Consulting Services. "The Inuit Interest in Seals and Sealing — Notes For A Brief to the Royal Commission." Presented by John Amagoalik and Peter Ernek. Montreal, January 1985.

Ronald, Keith and J.L. Dougan. "The Ice Lover: Biology of the Harp Seal (*Phoca groenlandica*). *Science*, Vol. 215 (19 February 1982).

Ronald, Keith, Jane Selley and Pamela Healey. "Seals: *Phocidae, Otariidae*, and *Odobenidae*." In *Wild Mammals of North America: Biology, Management and Economics*, pp. 769-827. Edited by J.A. Chapman and G.A. Feldhamer. Baltimore, Maryland: Johns Hopkins University Press, 1982.

Rowsell, H.C. "Report on Methods For Killing Seals." 1971. (Mimeographed)

Sergeant, David E. "Transatlantic Migration of a Harp Seal, *Pagophilus groenlandicus*." *Journal Fisheries Research Board of Canada*, Vol. 30, No. 1 (1973).

_____. "Environment and Reproduction in Seals." *Journal of Reproductive Fertility*, Suppl. 19 (1973), 555-61.

_____. "History and Present Status of Populations of Harp and Hooded Seals." *Biol. Conserv.*, Vol. 10, No. 2 (1976), 95-118.

Social Surveys (Gallup Poll) Limited. "Harpseal Study." London: September 1982.

Stewart, R.E.A. and Dave M. Lavigne. "Energetics of Nursing in the Harp Seal, *Phoca Groenlandica*, Energy Transfer and Female Condition." NAFO Scr. Doc. No. 81/XI/160 Serial No. N468. November 1981.

United States Code Annotated Title 16 Conservation No. 832 to End. "Cumulative Annual Pocket Part for use in 1981. Includes the Laws of the 96th Congress, Second Session (1980) through Public Law 96-486. Chapter 31 — Marine Mammal Protection." West Publishing Co., 1981.

United States of America. Public Law 97-58, 97th Congress. "An Act to improve the operation of the Marine Mammal Protection Act of 1972, and for other purposes." October 1982.

Walsh, John C. "Sealing in the 'Front' 1981 Season — Labrador Coast. A Report by John C. Walsh, Regional Director, World Society For The Protection of Animals." March 1981.

World Wildlife Fund. "Position on the 1982 Seal Hunt." March 1982.

Worthy, G.A.J. and D.M. Lavigne. "Changes in blood properties of fasting and feeding harp seal pups, *Phoca groenlandica*, after weaning." NAFO Scr. Doc. No. 81/XI/159 Serial No. N467. November 1981.

Printed in Canada